DISCARDED

599.0334
Ep4

126991

| DATE DUE | | | |
|---|---|---|---|
| | | | |
| | | | |
| | | | |
| | | | |
| | | | |
| | | | |
| | | | |
| | | | |
| | | | |
| | | | |
| | | | |

# EPITHELIAL-MESENCHYMAL INTERACTIONS IN DEVELOPMENT

# EPITHELIAL-MESENCHYMAL INTERACTIONS IN DEVELOPMENT

Edited by
Roger H. Sawyer
John F. Fallon

CARL A. RUDISILL LIBRARY
LENOIR RHYNE COLLEGE

## PRAEGER

PRAEGER SPECIAL STUDIES • PRAEGER SCIENTIFIC

**Library of Congress Cataloging in Publication Data**
Main entry under title:

Epithelial-mesenchymal interactions in development.

Bibliography: p.
includes index.
1. Epithelium. 2. Mesenchyme. 3. Developmental
biology. I. Sawyer, Roger H. II. Fallon,
John F.
QM561.E64  1983          599'.0334              82-13160
ISBN 0-03-060326-9

The cover shows an artist's drawing of the developing chick leg bud
and the epithelial-mesenchymal interface of the leg bud apex in
cross-section. Drawing by Lucy Taylor.

*599.0334*
*Ep4*
*126991*
*Dec. 1983*

Published in 1983 by Praeger Publishers
CBS Educational and Professional Publishing
A Division of CBS, Inc.
521 Fifth Avenue, New York, New York 10175 U.S.A.

© 1983 by Praeger Publishers

*All rights reserved*

3456789 052 987654321
Printed in the United States of America

For

Phyllis and Frank

Genie and Bill

# PREFACE

The means by which cells and tissues communicate during embryonic development are of fundamental interest to biologists. In many cases, as epithelial and mesenchymal tissues come together during development, an interaction occurs leading to specific alterations of one or both of the tissues. This process is known as embryonic induction. Such interactions may be either permissive or instructive in nature, though the distinction between these two remains controversial among developmental biologists. Understanding these interactions well enough to resolve the controversy will require not only great care in the analysis of the developing systems studied, but also in the terminology used to describe what is observed.

The symposium on which this book is based was organized to promote an exchange of ideas among those interested in the role of epithelial-mesenchymal interactions in developing systems. The Introduction opens the book with a general discussion on permissive and instructive epithelial-mesenchymal interactions. In the first part of this book, Dr. Fallon presents an analysis of epithelial-mesenchymal interactions during limb development. Dr. Kollar presents the epithelial-mesenchymal interactions in tooth development as a model of instructive induction, while Dr. Cunha discusses the role of sex hormones in epithelial-stromal interactions of the genital tract. Dr. Reddi describes progress made on the role of the extracellular matrices in differentiation and morphogenesis, and Dr. Lillie discusses the use of defined extracellular matrices in regulating differentiation of epithelia.

The second half of the book opens with a discussion by Dr. Sawyer on the role of tissue interactions in morphogenesis of avian scales. Dr. Dhouailly then presents experimental studies on the effects of vitamin A on the avian integument. Next, Dr. Hardy analyzes her work on vitamin A in studies of mammalian skin differentiation and Dr. Hall presents information on the role of epithelial-mesenchymal interactions during the development of cartilage and bone. Finally, Dr. Maderson gives an evolutionary view of epithelial-mesenchymal interactions.

The editors of this book express their sincere thanks to the American Society of Zoologists for sponsoring the symposium on which this book is based and especially to Mary Wiley for organizing the meeting in Dallas, Texas. We also wish to express our thanks to the National Science Foundation and the National Institutes of Health for their financial assistance.

# LIST OF CONTRIBUTORS

G. R. Cunha

Department of Anatomy
University of California—San Francisco
San Francisco, California

D. Dhouailly

Laboratoire de Zoologie et Biologie
 animale
Universite Scientifique et medicale
 de Grenoble
Grenoble, France

J. F. Fallon

Department of Anatomy
University of Wisconsin
Madison, Wisconsin

J. M. Frederick

Cullen Eye Institute
Baylor College of Medicine
Houston, Texas

H. Fujii

Kumamoto University
Kumamoto City, Japan

B. K. Hall

Department of Biology
Dalhousie University
Halifax, Nova Scotia, Canada

M. H. Hardy

Department of Biomedical Sciences
University of Guelph
Guelph, Ontario, Canada

A. Jepsen

Tissue Culture Laboratory
Department of Oral Pathology
Royal Dental School
Aarhus, Denmark

E. J. Kollar

Department of Oral Biology
School of Dental Medicine
University of Connecticut Health Center
Farmington, Connecticut

J. H. Lillie

Department of Anatomy and Cell Biology
University of Michigan
Ann Arbor, Michigan

D. K. MacCallum — Department of Anatomy and Cell Biology, University of Michigan, Ann Arbor, Michigan

P. F. A. Maderson — Department of Biology, Brooklyn College of CUNY, Brooklyn, New York

B. A. Meloy — Department of Anatomy, University of Colorado, Health Sciences Center, Denver, Colorado

A. H. Reddi — National Institutes of Health, Bethesda, Maryland

D. A. Rowe — Department of Laboratory Medicine and Pathology, University of Minnesota, Minneapolis, Minnesota

R. H. Sawyer — Department of Biology, University of South Carolina, Columbia, South Carolina

P. Sengel — Laboratoire de Zoologie et Biologie animale, Universite Scientifique et medicale de Grenoble, Grenoble, France

J. M. Shannon — Department of Anatomy, University of Colorado, Health Sciences Center, Denver, Colorado

B. K. Simandl — Department of Anatomy, University of Wisconsin, Madison, Wisconsin

O. Taguchi — Department of Anatomy, University of Colorado, Health Sciences Center, Denver, Colorado

# CONTENTS

Preface                                                         vii

List of Contributors                                            ix

Introduction                                                    xiii

PART ONE

1.  Studies on Epithelial-Mesenchymal Interactions
    During Limb Development                                     3

    John F. Fallon, Donene A. Rowe, Jeanne M.
    Frederick, B. Kay Simandl

2.  Epithelial-Mesenchymal Interactions in the Mammalian
    Integument: Tooth Development as a Model for
    Instructive Induction                                       27

    Edward J. Kollar

3.  Epithelial-Mesenchymal Interactions in Hormone-
    Induced Development                                         51

    Gerald R. Cunha, John M. Shannon, Osamu Taguchi,
    Hirohiko Fujii, Beth A. Meloy

4.  Role of Extracellular Matrix in Cell Differentiation
    and Morphogenesis                                           75

    A. H. Reddi

5.  The Role of Defined Extracellular Matrices on the
    Growth and Differentiation of Mammalian Stratified
    Squamous Epithelium                                         93

    John H. Lillie, Donald K. MacCallum, Arne Jepsen

PART TWO

6. The Role of Epithelial-Mesenchymal Interactions in Regulating Gene Expression during Avian Scale Morphogenesis 115

   Roger H. Sawyer

7. Feather Forming Properties of the Foot Integument in Avian Embryos 147

   D. Dhouailly, P. Sengel

8. Vitamin A and the Epithelial-Mesenchymal Interactions in Skin Differentiation 163

   M. H. Hardy

9. Epithelial-Mesenchymal Interactions in Cartilage and Bone Development 189

   Brian K. Hall

10. An Evolutionary View of Epithelial-Mesenchymal Interactions 215

    Paul F. A. Maderson

INDEX 243

# INTRODUCTION

"Now to see a difficulty and to wonder at it, is to admit
one's ignorance."

Aristotle, Metaphysics

There is a long history of studies on tissue interactions during
normal embryonic development (for reviews see Deuchar, 1975;
Weiss, 1939; Wessells, 1977; Wolff, 1970). It is clear that changes
over developmental time including cell proliferations and morphogenetic movements of groups of cells, bring cells and tissues of differing developmental histories into association. In many cases, the
juxtaposition changes the developmental fate of the tissues that come
into contact with each other. Whether or not there is a necessary
relationship between such tissue interactions has been explored in
many ways, for example, by preventing contact in situ and by in
vitro manipulations. Such experiments have demonstrated embryonic
inductions for developing structures from all three germ layers. An
important type of tissue interaction occurs between epithelia and
mesenchyme.
  Over the past twenty years, several investigators have grappled
with the problems of terminology and definitions for embryonic tissue
interactions. We acknowledge our debt to Clifford Grobstein (1964,
1967), Howard Holtzer (1968), Antone Jacobson (1966), and Norman
Wessells (1977), among others, for the clarity of their thought in
this difficult area. Epithelial-mesenchymal interactions (or inductions)
may be defined as tissue interactions which lead to changes in one or
both tissues; these changes would not occur without the interaction.
Thus, if the ureteric bud fails to make contact with the metanephro-
genic mesenchyme, there will not be a kidney on that side of the
embryo. In this case, as in many others, there is a reciprocal interaction between the two tissues. One of the fundamental questions
raised by the numerous investigations on embryonic induction is
whether the inducing tissue imparts specific information to the responding tissue. The alternative is that the inducing tissue presents
a nonspecific stimulus which triggers the response in an already
determined group of cells. In a particularly insightful paper, Holtzer
(1968; see also Wessells, 1977) proposed the descriptors "instructive
and permissive" for the two possibilities. Instructive inductions result in the cells of the tissue(s) developing patterns and undergoing
syntheses that they would not otherwise have followed, thus the deter-

mination of the responding tissue is set by the specific tissue interaction. Permissive inductions result in the realization of the prospective fate of already determined tissues through the tissue interactions.

Wessells (1977) has refined some working criteria that are
helpful in distinguishing between instructive and permissive inductive
events.

1) Because of the interactions, the responding tissue develops
in a predictable way (instructive and permissive).

2) When the tissues are separated, they do not develop in the
expected fashion (instructive and permissive).

3) A particular inducer can change the prospective fate of a
tissue it would not normally be associated with; this must be specific
for the inducer (instructive).

When considering the data from experiments on embryonic
inductions, the developmental history of the interacting tissues is of
prime importance. A negative result relates only to that particular
developmental time. Earlier or later developmental stages could
produce different results, thus both of the interacting tissues must
be competent. Finally, tissues respond to inductive stimuli within
the limits of their own genetic information.

Apparent instructive interactions are exemplified by studies
carried out on the developing chick epithelium (reviewed by
McLoughlin, 1968; Saunders, 1958; Sengel, 1976). In these cases,
the epidermal derivatives formed, or the structure and differentiation
of the epithelium formed, is dependent on the underlying mesenchymal
component. Saunders, McLoughlin, and others have demonstrated
this using a constant source of ectoderm, while varying the source
of the mesenchyme. Several examples of tissue combinations done
in ovo are listed below:

Wing Bud Ectoderm
  + Wing Bud (dermal) Mesoderm → Wing Feathers
Wing Bud Ectoderm
  + Thigh (dermal) Mesoderm      → Thigh Feathers
Wing Bud Ectoderm
  + Foot (dermal) Mesoderm       → Scales and Claws

Similar combinations can be accomplished in culture; several examples are listed below:

Leg Bud Ectoderm Alone   → Keratinizes
Leg Bud Ectoderm
  + Gizzard Mesenchyme    → Mucous Epithelium with Cilia
Leg Bud Ectoderm               Endothelial-like
  + Myocardial Myoblasts  →   Simple Squamous Epithelium

These examples indicate the role of the mesenchyme in specifying a particular response in the epithelium and also indicate the repertoire of the epithelium. We are at a point in the analyses of epithelial-mesenchymal interactions where numerous descriptions have been carefully documented in the literature. The time is ripe to take this information and apply the tools evolved by molecular biology, biochemistry, genetics, and immunology to the problems of embryonic induction. In fact, such analyses have already started. In this volume, we present diverse approaches from some of the laboratories carrying out research in this fascinating area of developmental biology. It is the hope of the editors and the publishers that this work will attract new students to the field and stimulate new approaches to the problems that have been with us since the beginning of this century.

## REFERENCES

Deuchar, E. M. (1975). Cellular Interactions in Animal Development. Chapman and Hall, London.

Grobstein, C. (1964). Cytodifferentiation and its controls. Science 143:643-650.

Grobstein, C. (1967). Mechanisms of organogenetic tissue interaction. Natl. Cancer Inst. Monogr. 26:279-294.

Holtzer, H. (1968). Induction of chondrogenesis: A concept in quest of mechanisms. In Epithelial-Mesenchymal Interactions (R. Fleischmajer and R. Billingham, eds.), pp. 152-164. Williams & Wilkins, Baltimore.

Jacobson, A. G. (1966). Inductive processes in embryonic development. Science 152:25-34.

McLoughlin, C. M. (1968). Interaction of epidermis with various types of foreign mesenchyme. In Epithelial-Mesenchymal Interactions (R. Fleischmajer and R. Billingham, eds.), pp. 244-251. Williams & Wilkins, Baltimore.

Saunders, J. W., Jr. (1958). Inductive specificity in the origin of integumentary derivatives in the fowl. In A Symposium on the Chemical Basis of Development (W. D. McElroy and B. Glass, eds.), pp. 239-254. The Johns Hopkins Press, Baltimore.

Sengel, P. (1976). Morphogenesis of Skin. Cambridge University Press, Cambridge.

Weiss, P. A. (1939). Principles of Development. Holt, New York.

Wessells, N. K. (1977). Tissue Interactions and Development. W. A. Benjamin, Menlo Park.

Wolff, E. (1970). Tissue Interactions During Organogenesis. (E. Wolff, ed.), Gordon and Breach Scientific Publishers, New York.

# EPITHELIAL-
# MESENCHYMAL
# INTERACTIONS IN
# DEVELOPMENT

# PART ONE

# 1
# STUDIES ON EPITHELIAL-MESENCHYMAL INTERACTIONS DURING LIMB DEVELOPMENT
## John F. Fallon, Donene A. Rowe, Jeanne M. Frederick, and B. Kay Simandl

## INTRODUCTION

Since the seminal works of Saunders (1948) and Zwilling (1955), the developing limb has been used to study epithelial-mesenchymal interactions (reviewed in Zwilling, 1961; Saunders, 1977). In this chapter, we will discuss work from our own and other laboratories on these interactions during limb development. Because of ready availability and relative ease of manipulation, many studies on limb development, especially those related to epithelial-mesenchymal interactions, have been carried out on the chick embryo wing bud. The generalizations drawn from the chick wing bud studies have been inferred to be true for other amniotes. Observations on the effects of mutations affecting the limb in mammals (Milaire, 1965) and teratogenic studies (Kameyama et al., 1973; Forsthoefel and Williams, 1975) are consistent with the experimental work done on the chick.

Tetrapod limbs begin development as thickenings of the mesoderm of the somatopleure, which soon take on a bud-shaped appearance. The thickening of mesoderm can be seen in the chick embryo at stage 17 (Hamburger and Hamilton, 1951) opposite somites 15-20 for the wing, and slightly later during stage 18 opposite somites 27-32 for the leg. The mesoderm of the limb bud is initially covered with typical embryonic epithelium, that is, simple cuboidal with overlying periderm. A change occurs in the epithelium at the apex of the wing bud during late stage 17 and stage 18 when the apical epithelium begins to thicken. By late stage 18 to early stage 19, the wing bud apex is covered by a pseudostratified columnar epithelium and overlying periderm termed the apical ectodermal ridge (Saunders,

1948). An apical ridge forms slightly later on the chick leg bud. All amniote limb buds have an apical ridge. Among reptiles and birds, the ridge takes the form of a pseudostratified columnar epithelium with periderm; among mammals the apical ridge is a stratified cuboidal to squamous epithelium with periderm (Fallon and Kelley, 1977).

## THE ROLE OF THE APICAL ECTODERMAL RIDGE IN LIMB DEVELOPMENT

In 1948, Saunders demonstrated that the apical ectodermal ridge is necessary for normal wing development (see also Summerbell, 1974). He showed that surgical removal of the apical ridge from the wing bud resulted in truncation of the developed wing. If the ridge was removed during stage 18, only the most proximal part of the humerus formed. Progressively more distal parts formed when the ridge was removed at later times, thus removal at stage 20 produced truncation at the level of the wrist. From these data and carbon marking experiments, it became clear that the limb elements differentiate in a proximo-distal sequence and the apical ridge is necessary for this development to occur. The experiments also showed that once removed the apical ridge could not regenerate.

Similar experiments with the same results have been done on the chick leg bud (Rowe and Fallon, 1982a) and on other bird embryos as well, e.g., the guinea fowl (Fallon, unpublished).

A more immediate effect of apical ridge removal at stages 18-20 is a pattern of cell death (Fig. 1.1) in the subridge mesoderm (Rowe, Cairns, and Fallon, 1982). Five hours after ridge removal, cells began to die in the anterior distal mesoderm. Cell death subsequently occurred in the posterior subridge mesoderm. The number of dying cells increased to reach a maximum by 8 hours after ridge removal when a band of necrosis extended nearly the antero-posterior length of the limb bud and 150-200 $\mu$m deep into the bud. By 12 hours, the number of dying cells diminished and at 16 hours cell death was completely absent. At this same time, the epithelium had healed over the wound; however, no new ridge formed. When the ridge was removed and replaced immediately, no cell death was observed. Removal of dorsal ectoderm leaving the ridge intact resulted in some death due to trauma but did not result in the pattern or massive extent of cell death seen when the ridge was removed.

In contrast, when the apical ridge was removed from stages 22-23 wing buds, the pattern of cell death shown in Figure 1.1 was not observed and fewer cells died. Ridge removal at even later developmental times (stages 24-25) resulted in no cell death in the

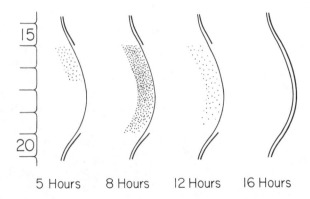

5 Hours    8 Hours    12 Hours    16 Hours

Figure 1.1 This drawing illustrates the pattern of cell death in the mesoderm of the stage 19-20 chick wing bud between 5 and 12 hours after surgical removal of the apical ectodermal ridge. The stippling indicates the areas of cell death. The unbroken line at 16 hours indicates healing of the apical epithelium, but no new apical ridge formed. The blocks on the left represent the wing bud level somites, 15-20.

subridge mesoderm. There appears to be a maturation of the subridge limb bud mesoderm from dependence on the apical ridge for survival.

The answer to the question of whether the mesodermal cell death seen after ridge removal is related directly to the truncation of the formed limb is not straightforward. It has been observed that after ridge removal, there are invariably some living cells (as many as 3 or 4 cell layers) at the wound surface when the deeper mesoderm cells die. This leads to the possibility that the few cell layers immediately beneath the ridge can survive ridge removal (cf. Saunders, 1977; Iten, 1982). These may be the cells that should give rise to the distal limb elements (Stark and Searls, 1973). However, if the ridge plus 100 $\mu$m of subridge mesoderm were removed, cell death was seen in the same spatio-temporal pattern as described above for simple ridge removal, except cells died only 100 $\mu$m into the mesoderm. The important point for this discussion is that living cells again were observed at the wound surface while deeper mesodermal cells died (Rowe et al., 1982). It is probably not correct to ascribe special properties of survival to the mesodermal cells immediately beneath the ridge. It is not clear where the living cells come from in either case described above, but it is likely they are part of normal wound healing. Thus, the possibility remains that most, if not all of the cells that should give rise to distal elements are eliminated by cell death. But this is probably much too simple

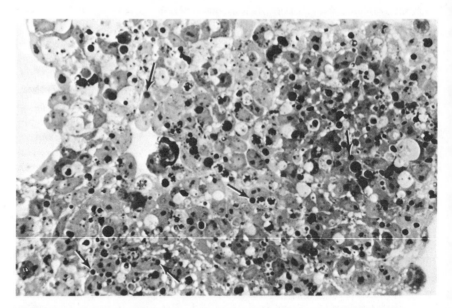

Figure 1.2 Histological section through an isolated apical ridge from a stage 19 wing bud grown in culture for 18 hours. The majority of the cells have died. The dark spots (arrows) are pycnotic nuclei of degenerating cells. A limb bud tip, apical ridge, and subjacent mesoderm, remained healthy for at least 24 hours when grown under the same culture conditions. (x 688)

a view, for, as already noted, cell death does not occur in the sub-ridge mesoderm after ridge removal after stage 23. In the wing, the entire manus, and in the leg the entire foot, was missing when the ridge was removed at stage 24. Obviously, cell death cannot be invoked to explain the failure of these structures to form after ridge removal. We suggest the ridge is required for survival of subridge limb mesoderm cells and for axial elongation at early limb bud stages. Subsequently (after stage 23), the mesoderm cells become independent of the ridge for survival but the ridge continues to be required for axial elongation through stage 28 in the wing (Summerbell, 1974) and 29 in the leg (Rowe and Fallon, 1982a).

The ridge cells in turn require the underlying limb mesoderm to survive. Searls and Zwilling (1964) have reported that the stage 19-20 apical ridge cells die within 24 hours when separated from limb mesoderm and grown in culture. We have recently confirmed these observations (Fig. 1.2; Boutin and Fallon, unpublished) and are now extending the experiments to later stages of development. It is of interest that the requirement of ridge cells for subridge mesoderm appears to be tissue specific. When other mesoderm was placed beneath the ridge, e.g., somite mesoderm, the ridge lost the pseudostratified columnar morphology and ridge cells died (Searls and Zwilling, 1964; Fallon, unpublished). These data seem to argue for a specific interaction between the limb mesoderm and the ectodermal ridge. Because limb bud mesoderm cannot induce a ridge after stage 17, this mesodermal activity would be different from the original ridge induction and similar to the much maligned "maintenance factor" proposed by Saunders and Zwilling (reviewed in Zwilling, 1961; Saunders and Gasseling, 1968).

In the stage 18-20 limb, the epithelial-mesenchymal components require association for normal differentiation, pattern formation, and morphogenesis, and also for survival of the cells of both tissues.


RIDGE INDUCTION: AN INSTRUCTIVE EVENT?

Several laboratories have demonstrated that the apical ridge of the chick limb bud is induced to form by the limb mesoderm. Saunders and Reuss (1974) as well as Dhouailly and Kieny (1972) have shown (using different techniques) that prospective wing bud mesoderm from stages 11 to 17, but not later, will induce an apical ridge in flank ectoderm. The data are not as precise for the host flank, but it appears the flank ectoderm is competent to respond to the prospective limb mesoderm by forming a ridge between stages 11 and at least through stage 17, but not later than stage 19. In both laboratories, the protocol was designed so that flank ectoderm healed over the implanted limb mesoderm. An apical ridge then developed in the flank ectoderm over the forming supernumerary limb bud (see Fig. 1.3). Competent limb mesoderm implanted in a pocket of flank so that flank epithelium did not have to heal over it did not induce a ridge and no limb formed (Saunders and Reuss, 1974).

It has been reported that early removal of the wing or leg apical epithelium (in a stage 17 wing; stage 17, 18 leg) will result in normal limb development in a low percentage of cases (Kieny, 1968; Fraser and Abbott, 1971; Saunders and Reuss, 1974; Rowe and Fallon, 1982a). It has been proposed (e.g., Kieny, 1968) that the conditions for apical ridge induction were still present in these few cases, and in the presence of the healed epithelium a new ridge was induced. Why wound healing is one of the conditions necessary for ridge induction under experimental conditions is not understood. Certainly this is not true in normal ridge induction.

Recently we have grafted stage 15 quail flank ectoderm (separated from the mesoderm with EDTA [disodium salt of ethylenediaminetetraacetic acid] treatment) to the wing area of the stage 15

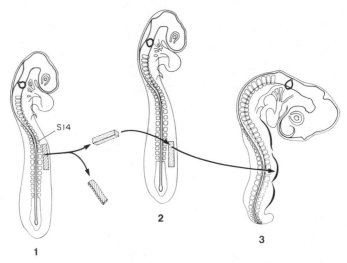

Figure 1.3 This drawing illustrates the procedure for grafting prelimb bud mesoderm to the flank. A donor stage 15 embryo (1) from
which the prospective limb region will be removed is shown on the
left. The 14th somite is indicated by S14. The donor mesoderm
(stippled) is separated from the ectoderm (parallel lines) by trypsin
or EDTA treatment. The ectoderm is discarded and the mesoderm
grafted to the flank of a host embryo (2). This may be done by rolling
the graft up and inserting it into a slit made in the flank (Dhouailly and
Kieny, 1972) or by making a tunnel in the flank mesoderm and inserting the graft beneath the flank ectoderm. The second procedure requires that the overlying flank ectoderm be scored or poked (wounded)
with a sharp instrument, or outgrowth will not occur (Saunders and
Reuss, 1974). In both cases, the flank heals over the graft of prospective limb mesoderm. Subsequently, the flank forms an apical
ridge (dark line) on the developing supernumerary limb bud on the
host embryo (3) flank. Developing limb buds of the host with overlying
apical ridge (dark lines) are shown in their normal positions. This
figure is based on Saunders' Patterns and Principles of Animal Development (Macmillan, New York, 1970) and has been modified and
redrawn.

chick. The host chick ectoderm was removed before the graft was
placed. A normal limb formed in the embryos in a high percentage
of cases. These were covered by quail ectoderm and the ridge that
was induced was made of quail cells (Fig. 1.4; Carrington and Fallon,
unpublished). There may be processes similar to wound healing in
these experiments also. Nevertheless, it would appear that specifically prospective limb mesoderm has, for a limited time, the

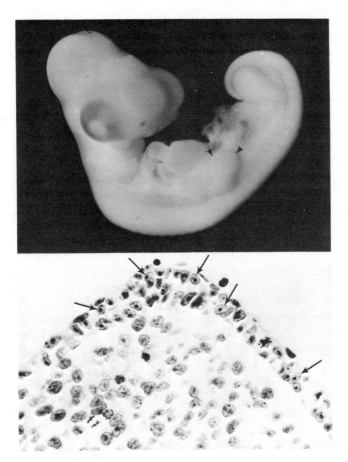

Figure 1.4  At stage 15, the wing level ectoderm was removed from the embryo pictured at the top of this plate and quail flank ectoderm was grafted in its place. A normal wing developed (arrowheads point to developing wing). The micrograph at the bottom of this plate is a cross section (x 609) through the developing wing, stained by the Feulgen method, and shows that an apical ridge was induced in the quail flank ectoderm by the chick wing mesoderm. The quail cells (arrows) are easily distinguished from the chick cells (LeDouarin and Barq, 1969) on the basis of the large heterochromatin clumps in the nuclei.

capacity to induce in competent ectoderm a morphological pattern, namely, pseudostratified columnar epithelium. This epithelium in turn has the capacity to induce limb outgrowth. It is reasonable to assume that the apical ridge acts in some way on the subjacent mesoderm. During the limb bud stages of development, flank ectoderm cannot support limb bud outgrowth; therefore, it is likely that when a ridge is induced, the prospective limb mesoderm also induces patterns of synthesis in the flank ectoderm that these cells would not normally possess. Far more work is necessary to assess what "ridge induction" means; however, at present it appears to be an instructive epithelial-mesenchymal interaction as defined in the Introduction to this volume.

## APICAL RIDGE ACTION ON LIMB MESODERM IS PERMISSIVE

The question may now be raised whether the interaction that occurs between the ectodermal ridge and the subjacent mesoderm is

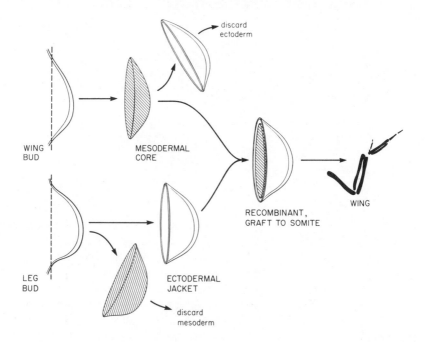

Figure 1.5 This drawing illustrates the procedure for making recombinant limbs. Wing and leg buds are removed from the embryo, ectoderm and mesoderm are separated with trypsin or EDTA. The mesodermal cores and ectodermal jackets may be combined in the desired combinations and grafted to a host embryo. In the case shown, a wing forms because the mesoderm was of wing level origin. Had there been leg mesoderm contamination in the leg bud ectodermal jacket, toes would have formed in the recombinant limb indicating the contamination. Thus, recombinants set up as shown have a built-in control to assess mesodermal contamination of the ectodermal jacket.

permissive or instructive. Two sets of experiments clearly indicate the answer. Zwilling (1955) devised a powerful technique to make what have been called recombinant limbs (Fig. 1.5). Isolated limb buds were treated with either trypsin or EDTA, permitting removal of the intact ectoderm from the mesoderm, which also remained intact. A recombinant limb of wing mesoderm with leg ectoderm formed a wing, while the reverse combination resulted in a leg. Therefore, it was clear that whether a forelimb or hindlimb formed was dependent on the source of the mesoderm (Zwilling, 1956).

Rubin and Saunders (1972) used the recombinant technique to demonstrate that heterochronic recombinant limbs were normal. The problem addressed was whether ridge action on the mesoderm had proximo-distal level specificity. It was found that if a stage 19 mesoderm was recombined with a stage 24 ectoderm, the limb that formed was normal. The reverse combination also produced normal limbs. Had the ridge been supplying level specific information about the proximo-distal axis, limb elements should have been missing along this axis in the first case and additional elements present in the second.

These two sets of elegant experiments clearly demonstrate the permissive nature of the apical ridge action on the limb mesoderm. In a way not understood, the ridge creates an environment which permits outgrowth of the underlying mesoderm cells without specifying the type of limb formed (fore- or hindlimb) or anything about the proximo-distal pattern of the limb elements.

## EXPERIMENTS WITH A LIMBLESS MUTANT

We have carried out a series of experiments (Fallon, Frederick, and Simandl, in preparation) using a line of chickens carrying a simple Mendelian gene which, in the homozygous condition, have no fore- or hindlimbs (Prahlad et al., 1979), while the heterozygotes are normal. The limbless condition is lethal because the chick cannot hatch. However, if hatched by hand, the chick can live for at least several weeks without apparent problems (Prahlad et al., 1979; McGibbon and Fallon, unpublished). At stage 17 in limbless embryos, a normal appearing wing bud does develop. However, the ectoderm does not form an apical ridge and by mid- to late stage 18, the wing bud, and slightly later the leg bud (Fig. 1.6) develop a characteristic depression in the middle of the bud. Cell death begins in the mesoderm at this time and the limb buds disappear completely by stage 24.

In order to determine which germ layer was affected by the limbless gene, we constructed recombinant limbs (cf. Fig. 1.5) from stage 19 and 20 normal and limbless limb bud components in the following fashion:

Limbless (leg or wing) mesoderm
+ normal ectoderm               → Normal leg or wing
Normal (leg or wing) mesoderm
+ limbless ectoderm               → No development

It is clear that in the limb bud stages of development, the limbless mesoderm is normal in that when supplied with a normal ectodermal jacket, it will form a limb that is normal in all respects. The
ectoderm is the affected germ layer and does not support growth of
the mesoderm. Experiments are in progress to learn whether the
limbless prelimb bud mesoderm cannot induce an apical ridge or
whether the limbless ectoderm cannot respond to the induction.

These experiments with the limbless embryo demonstrate that
limb bud outgrowth can be <u>initiated</u> without the apical ectodermal

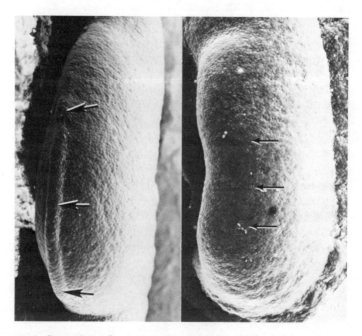

Figure 1.6 Scanning electron micrographs comparing the normal
stage 22 leg bud on the left, (x 96) with the stage 22 limbless leg bud
on the right, (x 84). Large arrows indicate the apical ectodermal
ridge on the normal leg bud. Small arrows indicate the characteristic depression of the saddle shaped limbless leg bud; no apical ridge
is present. Histological sections show cells are dying in the limbless
mesoderm at this time. However, a recombinant limb made of limbless mesoderm of this stage with a normal wing ectodermal jacket
will form a normal leg.

ridge. However, at the point when the morphologically identifiable ridge should appear, the limbless limb bud mesoderm cells began to die. We infer from this that the ridge is necessary to sustain outgrowth of the limb bud after its initial development. Without ridge development and function, the limb mesodermal cells will die.

## THE STRUCTURE AND PHYSIOLOGY OF THE APICAL RIDGE

How does the apical ridge carry out its function? We know very little, if anything at all, of the answer to this question. One approach has been to examine the fine structure of the apical ridge with the hope of being able to make correlations between structure and function. It has been shown that numerous very large gap junctions are characteristic of all amniote apical ridges examined (Kelley and Fallon, 1976; 1981; Fallon and Kelley, 1977). This was not true of the dorsal and ventral limb bud epithelia, where gap junctions were very small and widely dispersed. Among the various species of birds examined, the large gap junctions were seen to be clustered at the base of the pseudostratified epithelium (Figs. 1.7 and 1.8) of the ridge. Gap junctions were present at higher levels of the epithelium also, but were far fewer in number than at the base. In contrast, in the stratified mammalian ridge (Figs. 1.9 and 1.10), the numerous, large gap junctions were not noticeably clustered, but showed a relatively uniform distribution throughout. These distributions led us (Fallon and Kelley, 1977; Kelley and Fallon, 1981) to propose that the apical ridge cells are coupled via the gap junctions.

A direct test of this was accomplished (Fallon, Sheridan, and Clark, in preparation) by injecting the fluorescent dye Lucifer yellow CH into single cells of the apical ridge of chick wing and leg buds and examining the ridge for dye transfer. Dye was seen to move to adjacent cells almost immediately after injection, and then to very rapidly fill the cells of the entire ridge (Fig. 1.11). Dye was not observed to pass to the mesoderm nor to the dorsal or ventral ectoderms. Histological sections of injected ridges showed the dye did not pass to the periderm, which is consistent with the lack of gap junctions between ridge and periderm cells.

Thus, the cells of the apical ectodermal ridge are in communication with each other. Whether this has anything to do with ridge function is not clear at this point. Experiments are in progress to test this possibility.

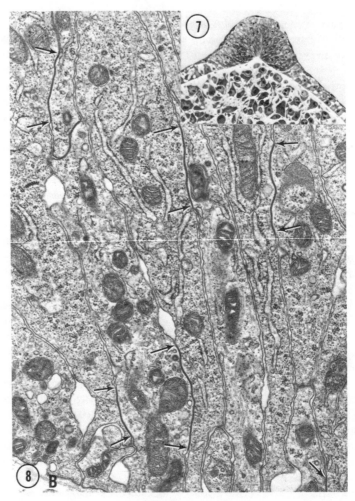

Figure 1.7  Light micrograph of the avian apical ridge in cross section (chick leg bud at stage 23) showing the typical pseudostratified columnar epithelium. (x 270)

Figure 1.8  Transmission electron micrograph from a section taken from stage 23 leg bud. Note basal lamina (B) at lower left of field. The micrograph shows the clustering of large gap junctions (arrows) at the base of the ridge. (x 20,243)

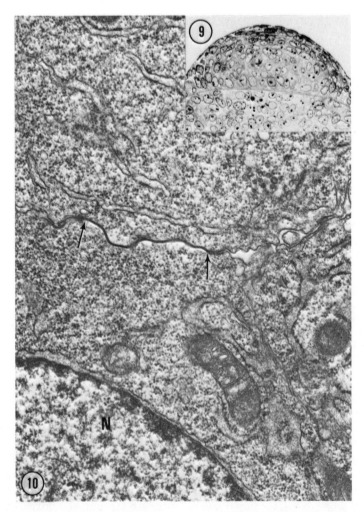

Figure 1.9 Light micrograph of the mammalian apical ridge in cross section (pig leg bud at 11 mm stage) showing the typical stratified squamous epithelium. (x 270)

Figure 1.10 Electron micrograph showing a typical gap junction (arrows) between two cells of a mammalian apical ridge. N marks the nucleus of the cell on the bottom of the micrograph. (x 27,512)

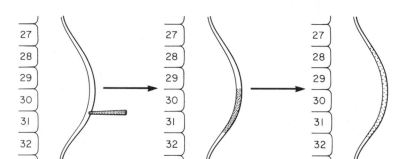

Figure 1.11 This drawing illustrates the spread of the fluorescent dye Lucifer yellow CH (indicated by the stippling) after injection with a pipette into a single leg bud apical ridge cell. The dye spreads rapidly, continues to spread after the pipette is removed, and stays within the apical ridge. Blocks on the left of each limb bud, numbered 27-32, represent leg bud level somites.

## EFFECT OF PARTIAL RIDGE REMOVAL ON RIDGE FUNCTION

A variety of observations in other laboratories made us question whether the anterior portion of the wing bud ridge could function when separated from the posterior wing bud ridge (Rowe and Fallon, 1981). For example, Warren (1934) showed that the anterior half of the wing bud failed to develop when the posterior half was removed. However, the reverse was not true; the posterior half of the wing bud did develop in the absence of the anterior half (see also Amprino and Camosso, 1955). Even more striking are experiments by Summerbell (1979) who placed an impermeable barrier through the dorsal-ventral extent of the stage 20 wing bud including the ridge. Structures, specifically the digits, expected to develop anterior to the barrier failed to develop (see also Kaprio, 1981). Digits expected to develop posterior to the barrier did form. When the barrier was placed through the dorsal-ventral extent of the mesoderm only without transecting the ridge, both anterior and posterior structures formed normally.

Taken together, these data seemed to indicate a difference in the functional capacity of the anterior and posterior parts of the wing ridge. To test this hypothesis, we had to determine the anteroposterior level of ridge which must be present for each of the digits to grow out normally. This was accomplished for stages 19-20 wing buds in a series of embryos, by removing increasing amounts of ridge starting with anterior ridge and moving toward the posterior ridge. We observed that when ridge was removed from opposite the middle of somite 15 posteriorly to the region opposite the junction between somite 17 and 18, all three wing digits formed. In other embryos, when the ridge removal was continued posteriorly to the ridge opposite the middle of somite 18, digit 2 failed to develop while digit 3 and 4 developed normally. Proceeding in this fashion, increasing the length of ridge removed by one-half somite amounts, anterior to posterior, the ridge was mapped. Figure 1.12 shows the maximum amount of ridge responsible for each digit in the wing at stages 19-20. On the basis of this information, the effect of removing posterior ridge on the function of the anterior ridge could be assessed. Complementary examples of anterior to posterior and posterior to anterior ridge removals are shown in Figures 1.13-1.16. Anterior to posterior removal to the level of mid-somite 18 (inset, Fig. 1.13) eliminated development of digit 2 (Fig. 1.13). In the complementary experiment where posterior to anterior ridge was removed (inset, Fig. 1.14), the ridge for digit 2 remained in place, yet digit 2 failed to develop. Similarly when ridge was removed from anterior to posterior (inset, Fig. 1.15) to the level opposite the junction of somites 18 and 19, digits 2 and 3 failed to develop (Fig. 1.15). In the complementary experiment, posterior to anterior ridge was removed (inset, Fig. 1.16), leaving the ridge for digits 2 and 3 intact,

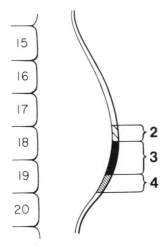

Figure 1.12 This drawing illustrates the maximum amount of apical ridge responsible for the outgrowth of the 3 digits of the chick wing at stage 19-20. The blocks on the left, numbered 15-20, represent the wing bud level somites (from Rowe and Fallon, 1981).

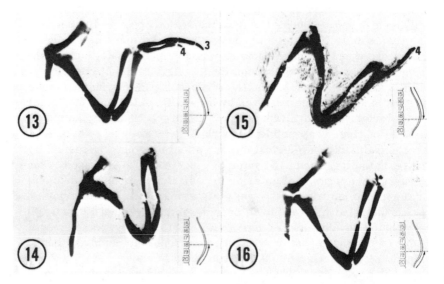

Figures 1.13 and 1.14 Complementary wing bud anterior-posterior and posterior-anterior apical ridge removals to the level of the middle of somite 18 are diagrammed in the inset drawings. The photograph illustrates that the limb formed after the anterior-posterior removal (Fig. 1.13) lacks digit 2, while digits 3 and 4 are normal. The wing formed after the complementary posterior-anterior ridge removal (Fig. 1.14) lacks digits 3 and 4 as expected, but also lacks digit 2 even though the ridge for digit 2 was not removed (from Rowe and Fallon, 1981).

Figures 1.15 and 1.16 Complementary wing bud anterior-posterior and posterior-anterior apical ridge removals to the level of the junction of somites 18 and 19 are diagrammed in the inset drawings. The photograph illustrates that the limb formed after the anterior-posterior removal (Fig. 1.15) lacks digits 2 and 3, and digit 4 is present. The wing formed after the complementary posterior-anterior ridge removal (Fig. 1.16) lacks digit 4 as expected, but also lacks digits 2 and 3 even though the ridge for these digits was not removed (from Rowe and Fallon, 1981).

yet these digits failed to develop (Fig. 1.16). Finally, we removed 150 μm ridge segments (half the length of a somite at these stages) from consecutive portions of the ridge on different embryos. We found that wing bud ridge anterior to the level opposite the middle of somite 19 cannot function (expected digits fail to develop) when separated from more posterior ridge.

There is an interesting fact about the avian wing bud that has been noted (Zwilling, 1961) but has attracted little attention over the years. The apical ridge displays an asymmetry from anterior to posterior on the wing bud tip. The ridge on the anterior (or preaxial) tip is not as high a pseudostratified columnar epithelium as that of the posterior (or postaxial) ridge. In fact, there is a difference of at least 10 μm in the height of the anterior compared to posterior ridge (Todt and Fallon, unpublished) at late stage 19. This is not the case in the avian leg bud, however, where the apical ridge is symmetrically disposed on the limb bud tip. It is worth stressing at this point that all mammalian limb buds examined to date (fore- and hindlimbs) have symmetrical apical ridges.

These observations led us to question whether anterior ridge of the chick leg bud would fail to function when posterior leg bud ridge was removed. By anterior to posterior ridge removals, we produced a map (Fig. 1.17) for the regions of the leg ridge responsible for each of the 4 leg digits (Rowe and Fallon, 1982b). Complementary anterior to posterior and posterior to anterior ridge removals are shown in Figures 1.18-1.21. Here we see that complementary ridge removals produced complementary digital patterns. When anterior ridge for particular digits was left after posterior to anterior ridge removal, the expected digits did form.

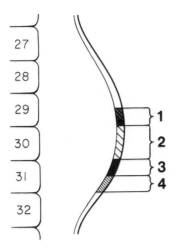

Figure 1.17 This drawing illustrates the maximum amount of apical ridge responsible for outgrowth of the 4 digits of the chick leg at stage 19-20. The blocks on the left numbered 27-32 represent the leg bud level somites (from Rowe and Fallon, 1982b).

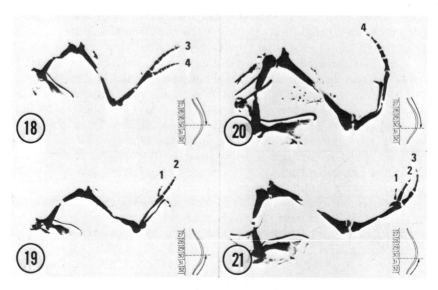

Figures 1.18 and 1.19 Complementary leg bud anterior-posterior and posterior-anterior apical ridge removals to the level of the middle of somite 30 are diagrammed in the inset drawings. The photograph illustrates that the limb formed after the anterior-posterior removal (Fig. 1.18) lacks digits 1 and 2. The leg that forms after the complementary posterior-anterior ridge removal (Fig. 1.19) lacks digits 3 and 4 as expected, but digits 1 and 2 are present.

Figures 1.20 and 1.21 Complementary leg bud anterior-posterior and posterior-anterior apical ridge removals to the level of the junction between somites 30 and 31 are diagrammed in the inset drawings. The photograph illustrates that the limb formed after the anterior-posterior removal (Fig. 1.20) lacks digits 1, 2, and 3. The leg that forms after the complementary posterior-anterior ridge removal (Fig. 1.21) lacks digit 4 as expected, but digits 1, 2, and 3 are present. Thus, complementary legs are produced from complementary apical ridge removals.

We have also placed impermeable barriers through the dorsal-ventral extent of the leg bud, including the apical ridge, in experiments similar to those reported by Summerbell (1979) for the wing bud. We found that leg structures expected to develop anterior to the barrier did develop (Rowe and Fallon, 1982b). In the leg then, anterior ridge will function when separated from posterior ridge.

Why does anterior wing bud ridge fail to function when posterior ridge is removed or anterior ridge is separated by an impermeable barrier, while this is not true in the leg bud? Obviously in the wing barrier experiments, the total amount of wing ridge potentially acting on the subjacent mesoderm is very nearly the same as in the normal wing bud. However, wing anterior ridge apparently cannot function in such circumstances. One possibility that can be tested is that the total coupled volume, via gap junctions, of the apical ridge is one important factor for its proper function. Thus, in the wing bud, removing (or separating) anterior ridge from posterior ridge results in a significant reduction in the total coupled volume of anterior ridge. This may be thought of in terms of a critical threshold level of coupled volume. In contrast, in the symmetrical leg ridge, removal of posterior ridge causes a comparatively insignificant loss of coupled volume, presumably above the threshold level for function. We infer limb buds of mammals, having symmetrical ridges, would behave in a way similar to the chick leg bud. Finally, all things considered, we feel it useful to be alert to the possibility that the avian leg, not the wing, may be a more suitable model for amniote limb development.

## POSSIBLE FUTURE APPROACHES TO THE PROBLEM

Obviously there are differences between the chick wing and leg bud apical ridges. They are different morphologically and they respond differently to the same experimental conditions. We raise the question whether all birds have the same dichotomy between fore- and hindlimb buds. There is the possibility that the observed dichotomy is an indication of part of the developmental changes that occurred during evolution, and produced the drastically modified forelimb that resulted in the ability to fly. Modern birds would have inherited these changes from their ancestors. Our work has been on the chick and it is important to note that Galliformes (which include the chicken) do not fly long distances, but explosively raise their bodies into the air with virtually no take-off run, and at the same time, are among the strongest fliers for short distances (Feduccia, 1980). Further, it is of interest that this order of birds probably has never given rise to any flightless species (Feduccia, 1980).

It would be of great interest to extend the observations we have begun to other orders of birds that use their wings very differently. Two likely candidates for comparison would be the order Apterygiformes (kiwis) that have almost no wing at all, and the order Gruiformes (the rails) which are the most likely of the extant birds to develop flightlessness (Feduccia, 1980). Similarly, soaring birds from the order Procellariiformes (albatrosses and shearwaters) should be compared with hovering birds from the order Apodiformes (hummingbirds and swifts). Wing buds from the order Spheniseiformes (penguins) also should be analyzed, not only to contrast them with other birds, but as a point of comparison with mammalian orders such as the Sirenia (manatees) and the Pinnipedia (seals, walruses) that may use their forelimbs in a similar manner. Finally, it would be useful to examine members of the mammalian order Chiroptera (bats) which also have achieved flight by means of drastically altered forelimbs.

CONCLUSION

The developing limb remains a superb system for analysis of epithelial-mesenchymal interactions. In this system, the induction of the apical ridge appears to meet the criteria of an instructive inter- action, while subsequent interactions between the ridge and meso- derm are permissive. In the early stages of limb development, there is a necessary relationship between the two tissues such that the cells of each die when the two are separated from each other. How the ridge is induced and maintained specifically by the limb mesoderm and how the ridge in turn creates the environment for mesodermal growth and differentiation are questions which must now be addressed directly.

ACKNOWLEDGEMENTS

This work was supported by NSF Grant #PCM79-03980 and NIH Grant #T32HD7118. We are grateful to Eugenie Boutin, Jill Carring- ton, Allen W. Clark, Mary Ellen McCarthy, Harland W. Mossman, David B. Slautterback, and William Todt for their constructive criti- cism of this manuscript. Special thanks are due to Allen Clark and David Slautterback for many helpful discussions of the work and what it means. We thank Ms. Lucy Taylor for making the drawings and Ms. Sue Leonard for typing the manuscript.

REFERENCES

Amprino, R. E. and Camosso, M. (1955). Ricerche sperimentali sulla morfogenesi degli arti nel polo. J. Exp. Zool. 129:453–493.

Dhouailly, D. and Kieny, M. (1972). The capacity of the flank somatic mesoderm of early bud embryos to participate in limb develop- ment. Dev. Biol. 28:162–172.

Fallon, J. F. and Kelley, R. O. (1977). The ultrastructural analysis of the apical ectodermal ridge during vertebrate limb morphogenesis. II. Gap junctions as distinctive ridge structures common to birds and mammals. J. Embryol. Exp. Morphol. 41:223-232.

Feduccia, A. (1980). The Age of Birds. Harvard University Press, Cambridge.

Fraser, R. A. and Abbott, U. (1971). Studies on limb morphogenesis. V. The expresssion of eudiplopodia and its experimental modification. J. Exp. Zool. 176:219-236.

Forsthoefel, P. F. and Williams, M. L. (1975). The effects of t-fluorouracil and 5-fluorodeoxyuridine used alone and in combination with normal nucleic acid precursors on development of mice in lines selected for low and high expression of Strong's luxoid gene. Teratology 11:1-20.

Hamburger, V. and Hamilton, H. (1951). A series of normal stages in the development of the chick embryo. J. Morphol. 88:49-92.

Iten, L. E. (1982). Pattern specification and pattern regulation in the embryonic chick limb bud. Am. Zool. 22:117-129.

Kameyama, Y., Hayasaka, I., and Hoshino, K. (1973). Morphogenesis of 5-fluorouracil induced polydactylism in mice. Teratology 8:95-96.

Kaprio, E. A. (1981). Transfilter evidence for a zone of polarizing activity participating in limb morphogenesis. J. Embryol. Exp. Morphol. 65:185-197.

Kelley, R. O. and Fallon, J. F. (1976). Ultrastructural analysis of the apical ectodermal ridge during vertebrate limb morphogenesis. I. The human forelimb with special reference to gap junctions. Dev. Biol. 51:241-256.

Kelley, R. O. and Fallon, J. F. (1981). The developing limb: an analysis of interacting cells and tissues in a model morphogenetic system. In Morphogenesis and Pattern Formation (T. G. Connelly, L. L. Brinkley, and B. M. Carlson, eds.), pp. 49-85. Raven Press, New York.

Kieny, M. (1968). Variation de la capacité inductrice du mésoderme et de la competence de l'ectoderme au cours de l'induction primaire du bourgeon de membre, chez l'embryon de poulet. Arch. Anat. Micr. Morph. Exp. 57:401-418.

LeDouarin, N. and Barq, G. (1969). Sur l'utilisation des cellules de la Caille japonaise comme "marquers biologiques" en embryologie expérimentale. C.R. Acad. Sci. Ser. (Paris) D269:1543-1546.

Milaire, J. (1965). Aspects of limb morphogenesis in mammals. In Organogenesis (R. L. DeHaan and H. Ursprung, eds.), pp. 283–300. Holt, Rinehart and Winston, New York.

Prahlad, K. V., Skala, G., Jones, D. G., and Briles, W. E. (1979). Limbless: a new genetic mutant in the chick. J. Exp. Zool. 209: 427–434.

Rowe, D. A., Cairns, J. M., and Fallon, J. F. (1982). Spatial and temporal patterns of cell death in limb bud mesoderm after apical ectodermal ridge removal. Dev. Biol. 93:83–91.

Rowe, D. A. and Fallon, J. F. (1981). The effect of removing posterior apical ectodermal ridge of the chick wing and leg on pattern formation. J. Embryol. Exp. Morphol. 65:(Supp.) 309–325.

Rowe, D. A. and Fallon, J. F. (1982a). The proximodistal determination of skeletal parts in the developing chick leg. J. Embryol. Exp. Morphol. 68:1–7.

Rowe, D. A. and Fallon, J. F. (1982b). Normal anterior pattern formation after barrier placement in the chick leg: further evidence on the action of polarizing zone. J. Embryol. Exp. Morphol. (in press).

Rubin, L. and Saunders, J. W., Jr. (1972). Ectodermal-mesodermal interactions in the growth of limb buds in the chick embryo: constancy and temporal limits of the ectodermal induction. Dev. Biol. 28:94–112.

Saunders, J. W., Jr. (1948). The proximo–distal sequence of origin of the parts of the chick wing and the role of the ectoderm. J. Exp. Zool. 108:363–403.

Saunders, J. W., Jr. (1977). The experimental analysis of chick limb bud development. In Vertebrate Limb and Somite Morphogenesis (D. A. Ede, J. R. Hinchliffe, and M. Balls, eds.), pp. 1–24. Cambridge University Press, Cambridge.

Saunders, J. W., Jr. and Gasseling, M. T. (1968). Ectodermal-mesenchymal interactions in the origin of limb symmetry. In Epithelial-Mesenchymal Interactions (R. Fleischmajer and R. E. Billingham, eds.), pp. 78–97. Williams and Wilkins, Baltimore.

Saunders, J. W., Jr. and Reuss, C. M. (1974). Inductive and axial properties of prospective wing bud mesoderm in the chick embryo. Dev. Biol. 38:41-50.

Searls, R. L. and Zwilling, E. (1964). Regeneration of the apical ectodermal ridge of the chick limb bud. Dev. Biol. 9:35-55.

Stark, R. J. and Searls, R. L. (1973). A description of chick wing bud development and a model of limb morphogenesis. Dev. Biol. 33:138-153.

Summerbell, D. (1974). A quantitative analysis of the effect of excision of the AER from the chick limb bud. J. Embryol. Exp. Morphol. 32:651-660.

Summerbell, D. (1979). The zone of polarizing activity: evidence for a role in normal chick limb morphogenesis. J. Embryol. Exp. Morphol. 50:217-233.

Warren, A. E. (1934). Experimental studies on the development of the wing in the embryo of Gallus domesticus. Am. J. Anat. 54: 449-485.

Zwilling, E. (1955). Ectoderm-mesoderm relationship in the development of the chick embryo limb bud. J. Exp. Zool. 128:423-441.

Zwilling, E. (1956). Reciprocal dependence of ectoderm and mesoderm during chick embryo limb development. Am. Nat. 90:257-265.

Zwilling, E. (1961). Limb morphogenesis. Adv. Morphog. 1:301-330.

# 2

# EPITHELIAL-MESENCHYMAL INTERACTIONS IN THE MAMMALIAN INTEGUMENT: TOOTH DEVELOPMENT AS A MODEL FOR INSTRUCTIVE INDUCTION

*Edward J. Kollar*

During the last several years many reviews of tooth development have appeared (Hay and Meier, 1978; Slavkin, 1978; Kollar and Lumsden, 1979; Slavkin et al., 1980; Kollar, 1981; Thesleff and Hurmerinta, 1981). These discussions reflect the research perspectives of the writers and each is very different from the others. Nonetheless, these reviews indicate the intensified interest in this experimental model. This appreciation of the importance of developing teeth as a model for studying tissue interactions is very different from the limited interest in this model which existed ten years ago when I reviewed this area in the context of the developing integument (Kollar, 1972). It is my intent, again, to relate recent work on tooth development to the broader context of skin morphogenesis with special reference to the nature of induction. Because the literature has been reviewed so frequently, I will not describe in detail the work already reviewed. Instead, the implications of these data will be discussed in a broad context, and our more recent approaches will be described.

ODONTOGENESIS: SOME GENERAL PRINCIPLES

In the developing integument, it is clear that the morphogenesis of skin and its derivatives depends on reciprocal interactions between the epithelium and the connective tissue stroma (Billingham and Silvers, 1968; Kollar, 1972; Sengel, 1976; Kollar, 1981). In fact, as will be discussed below, epithelia depend on mesenchyme not only for survival, but they are specified regionally by mesenchyme to express specialized structures (Rawles, 1963; Kollar and Baird, 1970a, 1970b; Dhouailly, this volume; Sawyer, this volume). That is, mesenchymal components can induce differentiated structures in labile, developmentally plastic epithelial cells.

Dental papillae are able to elicit tooth morphogenesis in a variety of experimental combinations in epithelia of ectopic origin. Similarly, in skin, dermal papillae induce hair follicles in glabrous epithelia from the plantar surface or the tongue. Thus, once specified as papillae, mesenchymal cells control epithelial histogenesis. Such histogenesis can be exquisitely detailed; the overall shape of the structures as well as many fine details of cytodifferentiation and pattern are controlled by mesenchymal components (Sengel, 1976; Kollar, 1981). In experimental tissue combinations, if the epithelium originates from other regions of the skin, or from younger stages of a rudiment, these tissue interactions begin de novo and repeat a prescribed and predictable sequence of development. Three generalized stages: initiation, morphodifferentiation, and cytodifferentiation, can be discerned (Kollar and Lumsden, 1979; Kollar, 1981). These stages are usually seen as an early budding of the epithelium, followed by epithelial invasions into the mesenchyme which expand to take on specific shapes (hair, vibrissa, incisor, molar; salivary and mammary glands), and, finally, cytodifferentiation.

Wolpert (1981) argues from his positional information hypothesis that the mesenchyme, having been specified as an inducer, then imparts positional information to the epithelium. This idea may be internally consistent with the hypothesis but it is difficult to interpret it in terms of patterning in the integument. There is no information that compels us to think that the mesenchyme of the head or the integument is the origin of the patterning signal. Our idea that in the dental arch and in the vibrissal pad on the snout the innervation specifies the sites of the individual papillae (Kollar and Lumsden, 1979; Kollar, 1981) did not resolve the question of possible preexisting specification in the local mesenchyme, in migrating neural crest cells, or in the early ectoderm itself. It is not impossible that the specification of the inductive mesenchyme is the result of earlier influences of the overlying ectoderm on the mesenchyme.

Actually, the stage of initiation during which dental papillae are specified in a segmentally patterned dental arch has not been investigated with the vigor applied to later stages. This early event could very well be the instructive signal upon which all later interactions depend (Kollar, 1981). Once the two components of these rudiments are established, however, they retain this information for extended periods of time despite experimental interference with the interaction (Kollar and Kerley, 1980). Complete tissue dissociation followed by extended cell culture will not destroy the capability for induction and morphogenesis. If the cells are recombined and grafted, the cells sort out and repeat the entire sequence of tooth morphogenesis. Similar comments might be made for the importance of the mesenchymal condensations that form during the development of hair and vibrissae (Oliver, personal communication).

There is reciprocality in subsequent interactions controlling tooth development. The morphodifferentiation of ectopic epithelia can be directed by dental papillae, but at later stages these same dental papillae require newly induced enamel organ epithelia to begin dental matrix secretion. Then, at still later stages, the onset of enamel protein secretion in the enamel organ is dependent on the presence of mineralized dentin matrix. In this way, the developing tooth is similar in its complex reciprocal interactions to the developing limb in which an instructive event establishes the limb anlage and then continues through a complex series of reciprocal interactions (Fallon, this volume).

## THE INFLUENCE OF THE EXTRACELLULAR MATRIX (ECM)

The involvement of ECM in these interactions has been demonstrated repeatedly in many developing tissues. Salivary gland (Grobstein, 1953; Grobstein and Cohen, 1965), mammary gland (Bernfield et al., 1973; Bernfield, 1981), kidney (Saxen et al., 1981; Ekblom et al., 1980) and teeth (Kollar, 1981; Thesleff and Hurmerinta, 1981) have been examined and the importance of collagen and glycosaminoglycans (GAG) have been demonstrated. The ECM is necessary for the early events of tooth morphogenesis and for stabilizing previous development.

The complexity of the ECM is illustrated in Figure 2.1. The molecular heterogeneity of the local extracellular environment may provide sufficient substrate information to account for local inductive influences of regional mesenchyme throughout the embryo. Interactions between the plasma membrane and cytoskeleton with this substrate may provide the inductive signal. Thus, a specific, diffusible, inducer molecule may not be necessary (Kollar, 1981). Alterations of the extracellular matrix (Kollar, 1981; Lillie, this volume; Reddi, this volume) dramatically alter cell behavior.

Once the early stages of tooth development have established definitive tooth germs, dramatic changes take place at the epithelial-mesenchymal interface. Immunofluorescent probes specific for the components of the basement membrane have been used extensively to demonstrate these dynamic alterations in ECM composition and structure. Thesleff and others (Lesot et al., 1978; Thesleff and Hurmerinta, 1981) have described an intact basement membrane during the early stages of development and have discussed the alteration

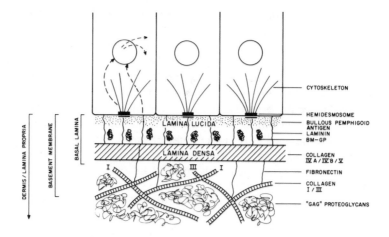

Figure 2.1 This is a hypothetical composite of the components of the ECM of the skin and oral mucosa. The collagens specific to the basal lamina (IV A, B; V) and to the adjacent stroma (I and III) are illustrated with respect to the non-collagenous glycoproteins of the ECM: laminin, fibronectin, glycosaminoglycans and proteoglycans as well as basement membrane glycoprotein (BM-GP) and bullous pemphigoid antigens. The interactions between the substrate and the cell may take place through unidentified surface receptors, the cytoskeleton, or both (closed arrowheads). In turn, via these transducing mechanisms the nucleus responds to the inductive ECM by transcribing specific genetic information characteristic of cytodifferentiation (open arrowheads). Absent from the figure are the fibroblasts or mesenchymal cells that may alter or control invasive patterns of the epithelium into the extracellular matrix. In addition, the specialized anchoring fibrils often found in the basement membrane region are not illustrated. (The figure is drawn from data discussed by Kleinman et al., 1981 and Briggaman, 1981.)

of fibronectin and GAG as dental papillae mesenchyme cells begin to align in preparation for deposition of type I collagen. Immediately following mineralization of the predentin, the basal lamina breaks down and the cells of the inner dental epithelium that were formerly basal cells associated with a basal lamina now differentiate into ameloblasts and secrete enamel proteins into this newly formed "lumen" (Kollar, 1981).

The disintegration of the basal lamina permits cell-cell associations to take place between the cells of the inner dental epithelium and the cell processes of differentiating odontoblasts in the dental mesenchyme (Kallenbach and Piesco, 1978). The significance, if any, of these cell-cell contacts during later stages of tooth development has not been established. In addition, these cell contacts are documented only in the final stages of cytodifferentiation after a great deal of morphogenesis has already taken place. Cell-cell associations have not been seen in the early stages of tooth morphogenesis during which early inductive events are taking place. This is in contrast to the developing kidney where the importance of cell-cell associations has been demonstrated (Saxen et al., 1980; Ekblom et al., 1980). In the tooth interactions with complex and manifold components of the ECM and basement membrane are likely to be the most important elements of the interaction (Kollar, 1981).

Transfilter studies preclude complex histogenesis in epithelium and mesenchyme because of physical constraints and, consequently, transfilter morphodifferentiation into teeth does not occur. Thus, transfilter studies which demonstrate the interaction of tissues with ECM can only be done with dental tissues that have experienced considerable prior developmental interaction. Thesleff et al. (1978) have demonstrated that, at later stages of tooth development, filters of optimal porosity and thickness can be interposed between the dental tissues. If the ECM components can traverse the filter, conditions that existed at the time of separation are reestablished and the cells undergo cytodifferentiation.

## INSTRUCTIVE VERSUS PERMISSIVE INTERACTIONS

Several aspects of tooth morphogenesis have had an important impact on our thinking about epithelial-mesenchymal interactions and much discussion has been focused on the nature of the inductive signal (Kollar, 1981; Fallon, this volume).

The induction of teeth in ectopic epithelium is, perhaps, the clearest example of an instructive tissue interaction. In this case, complex morphological structure is elicited in foreign epithelia by the action of inducing mesenchyme. Unique secretory products are deposited to form a complete tooth; new genetic information is transcribed in order to achieve a complex structure. Such interactions define instructive inductive interactions.

But, as a result of such instructive interactions, an enamel organ is induced and the dental papillae secrete type I collagen and other constituents of the dentin matrix. Later stages of tooth morphogenesis are absolutely dependent on the presence of a previously induced enamel organ. Thus, we have an example of a permissive tissue interaction at a later stage of tooth morphogenesis. It has not been possible to mimic this situation experimentally; a fully formed dental papilla will not secrete matrix in the absence of the enamel

TABLE 2.1

Tissue Interactions

| Structures | Source of | | Type of Induction | Epithelial Reaction to Heterologous Mesenchyme |
| | Epithelium | Mesenchyme | | |
| --- | --- | --- | --- | --- |
| Skin<br>Oral mucosa | Ectoderm | Mesoderm<br>Ectomesenchyme | Instructive | Responsive |
| Gut<br>Digestive glands<br>Lung | Endoderm | Mesoderm | Permissive | Refractory |
| Urogenital tract | Mesoderm<br>Endoderm | Mesoderm | Instructive<br>Permissive | Refractory |

Source: Compiled by author.

organ and no substitute epithelium has been found that will support dentin matrix deposition. On the other hand, it has not been possible experimentally to elicit dental papilla cells from ectopic cranial or somatic mesenchyme by using a fully formed enamel organ; the enamel organ does not act instructively. Thus, once tooth germs are induced, permissive interactions between enamel organs and dental papillae are bidirectional. Induced enamel organs interact with established dental papillae and provide the conditions necessary for matrix synthesis and deposition.

Tooth induction is clearly different from other tissue interactions such as pancreas in which a protodifferentiated cell type is potentiated by environmental (natural or experimental) conditions to increase dramatically the amounts of secretory products being synthesized. Nor should the tooth model be confused with essentially single-cell models such as cartilage or cornea in which a cell is permissively stimulated to produce large quantities of cell-associated ECM. Rather, in the experimental tooth combinations, new genetic information is expressed in foreign tissues at the end of a series of morphological transformations which utilize the new cell products structurally as a functional organ.

Furthermore, instructive inductions impart long-term stable properties to cells so that despite disruption of the tissue components, rudiments can be reconstituted experimentally. The requirements for stability in the inducer and for specific predictable morphological tissue responses seem to be characteristic features of tissue interactions in ectodermally derived structures. This is most clearly seen in the integument and also in certain structures of the urogenital tract (Cunha, this volume; Cunha et al., 1980).

Table 2.1 illustrates organ systems, epithelial origins, and inductive properties involved in various embryonic epithelial derivatives. Their origins in the early embryo seem to impart certain unique and stable characteristics and these can be roughly correlated with their germ layer origin. Thus, to lump all of these diverse models together in a discussion of the nature of tissue interactions is undiscriminating and consequently may be more confusing than synthetic.

Evolution has probably selected a variety of strategies to orchestrate the organization of embryos—some simple, some more complex—but each decidedly distinct. How much of cellular responses are restricted or preprogrammed into cells of various germ layer origins and how much is left to instructive interactions has yet to be ascertained. What is clear is that instructive induction is phylogenetically ancient and a general principle rather than a specific signal for each inductive event. There is evidence from a variety of experiments that inductive cues cross species and class lines (Dhouailly, 1973; and this volume; Coulombre and Coulombre, 1971; Fallon and Crosby, 1977; and Kollar and Fisher, 1980; Kollar, 1981) indicating the universality of this mechanism in vertebrates.

GLANDULAR VERSUS NONGLANDULAR
SKIN DERIVATIVES

The observation made earlier that during the last stages of tooth morphogenesis the basal lamina breaks down and enamel proteins are secreted into a "lumen" provided by this maneuver suggested a new series of experiments. The implication that former basal cells reverse their intracellular polarity and become glandlike secreting cells is an unusual transformation of cell function and morphology. The relationship and the associated tissue interactions between the glandular derivatives of the skin and oral mucosa (salivary glands, mammary gland, etc.) and the nonglandular adnexa (hair, vibrissae, teeth) is being investigated.

Salivary glands have played an important role in experimental studies of tissue interactions since the pioneering work of Grobstein (1953). These epithelial derivatives seem to be the logical starting place. What is immediately apparent is that the entire literature concerning salivary gland development has dealt with the branching pattern and the mesenchymal specificities concerned with this aspect of their morphodifferentiation (Lawson, 1972). Investigations of the questions of whether salivary gland mesenchyme can induce salivary glands in an ectopic epithelium, or whether salivary gland epithelium can be induced to form nonglandular structures have been neglected.

## PAPILLAE VERSUS CAPSULAR MESENCHYME

Essential differences in the way epithelia react to the mesenchyme, either as an invaginating, papilla-incorporating structure (teeth, hair, vibrissae), or as a branching, bifurcating, and lobule-producing structure surrounded by capsular mesenchyme have not been experimentally evaluated. There appear to be two distinct modes of epithelial-mesenchymal interactions operating in the integument. But how mesenchymal cells are specified as papillae or as capsular mesenchyme has not been considered.

There is an important demonstration of these distinctions in the work of Hardy and her colleagues (Hardy, 1968; Hardy and Bellows, 1978; Hardy et al., 1978 and Hardy, this volume) who have investigated the glandular metaplasia that occurs in embryonic mouse epidermis in response to vitamin A. Maintenance of stable and persistent glands after vitamin A treatment and the regenerative ability of the vibrissal follicles when control conditions are restored imply that vitamin A has altered the epithelial-mesenchymal interactions typical of this region of the embryo.

The stage-specific dependency, the ultrastructural and histochemical alterations of the mesenchyme, and the differential response when vitamin A is removed may indicate that permanent transformations of the developmental pathways of the epithelium and the mesenchyme have occurred. Hardy's description of the changes in the extracellular matrix suggests that normal dermal mesenchyme has taken on many of the characteristics of capsular mesenchyme. The important observations that this change is stable and that vibrissae can reappear offer new experimental possibilities.

For example, the mesenchyme associated with the glands should be tested for inductive potential in new epithelial-mesenchymal interactions with embryonic snout and glandular (salivary, mammary, etc.) epithelium. In addition, since the effect of vitamin A on tooth development has been tested only at later stages of development (Hurmerinta et al., 1980), this metaplasia-inducing effect of vitamin A should be examined on younger stages of tooth development comparable with the sensitive stages of vibrissae development. In this case the glandular bias of the vibrissae follicle as seen in its ability to produce sebaceous glands would be eliminated. While the tooth in later stages does secrete, it in no way assumes a branched glandular structure. Vitamin A-induced glandular metaplasia provides an experimental means of switching between glandular and nonglandular epidermal derivatives and may also provide the experimental means of changing dermal and papillar mesenchyme into capsular mesenchyme.

It is obvious that this discussion of the essential or intrinsic differences between glandular and nonglandular is a restatement of the old problem of the relationship between morphogenesis and cytodifferentiation. Does the morphology influence the expression of genetic information? Can the two developmental states be uncoupled? It has not been possible to uncouple the complex series of morphological events from cytodifferentiative events that occur in tooth development. During induction in experimental tissue combinations, the processes of morphodifferentiation are recapitulated before expression of specialized gene products are seen. If the two events are not causally related they are certainly linked operationally. Even in transfilter experiments, older tissues must be used and the influence of the ECM must be reestablished in order to permit secretion of the hard tissue matrices.

Previous attempts to examine this relationship between morphology and differentiation have dealt with reciprocal combinations of submandibular salivary glands and mammary gland tissues. Sakakura et al. (1976) and Kratochwil (1969) did not offer a definitive conclusion since, once again, considerable developmental interaction had taken place in these morphologically similar glandular tissues.

Experimental combinations made between glandular tissues and nonglandular derivatives such as the tooth germs seem to be a more definitive test of the influences of instructive tissue interactions. Salivary gland was chosen because it avoided the hormonal influences that complicate the developing mammary gland interactions. In addition, larger amounts of salivary gland tissue can be obtained at early stages of development. This issue as it relates to mammary gland development should also be examined in the context of our present findings, that these two structures may be developmentally determined by their germ layer history.

## RECOMBINATIONS OF SALIVARY GLAND AND DENTAL TISSUES

Isolation of epithelial and mesenchymal tissues of salivary gland and tooth germs was done by our routinely employed method, cold trypsinization (Rawles, 1963). The effectiveness of this method has been demonstrated before for tooth germs (Kollar and Baird, 1969, 1970a; Kollar, 1972) and can be seen in the scanning photomicrograph (SEM) in Figure 2.2A. Figure 2.2B illustrates a SEM of an early

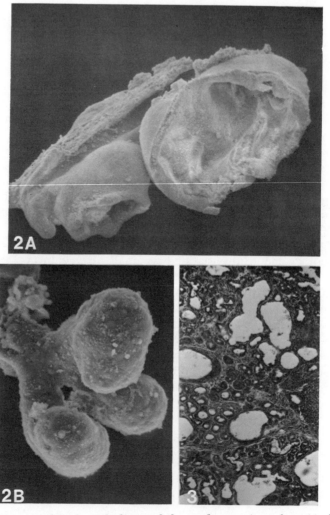

Figure 2.2A Isolated epithelium of the molar region of an 18-day-old mouse embryo. Enamel organs ($M_{1-3}$) are seen. SEM. (x 75)

Figure 2.2B Submandibular salivary gland epithelium isolated from a 14-day-old mouse embryo. SEM. (x 270)

Figure 2.3 Intraocular graft of a recombination of submandibular salivary gland epithelium and embryonic mouse molar papilla. Acini, ducts, and expanded cysts are present as well as bone. (x 75)

branching submandibular salivary gland epithelium from a 14-day mouse embryo. A similar picture can be obtained from 17-day parotid salivary gland rudiments. The epithelium is intact and devoid of mesenchymal cells. The mesenchyme was removed by mechanical dissection and not be flushing the tissue in a pipette as is often done with salivary gland tissue. Since salivary gland mesenchyme does not maintain its tissue integrity as readily as the tooth papilla and dental sac, the mechanical dissection must be done very gently or the salivary mesenchyme organization and many cells are lost. The tissue recombinants were explanted to the intraocular site for 1-2 weeks.

When submandibular gland epithelium is recombined with molar mesenchyme, dental induction does not occur (Fig. 2.3). Instead, glandular epithelium arranged as ducts, cysts, and acini are found. The molar mesenchyme differentiates as random spicules of bone. These data were disappointing and confirmed after many trials that submandibular epithelium is very stable and unresponsive to foreign mesenchyme influences. The data also confirm both Cunha's (1972) and Lawson's (1972) finding that the strict specificity of submandibular epithelium for its own mesenchyme is not absolute (Grobstein, 1953). Submandibular salivary epithelium is supported by molar mesenchyme. The data also indicate that the molar mesenchyme does not induce salivary epithelium to form teeth.

The reciprocal experiment of molar enamel organ epithelium associated with submandibular mesenchyme resulted in keratinizing cysts of epithelium with no evidence of tubule, duct, or branching acinar patterns derived from the dental epithelium. The range of tissue ages tested is limited and does not preclude an inductive influence of the salivary mesenchyme on a foreign ectopic epithelium in more favorable combination. These data were surprising in light of the previous discussion concerning the inductive properties and epithelial responsiveness of integumental derivatives (Table 2.1). Why does the submandibular tissue react differently? Are there intrinsic differences between glandular and nonglandular tissues? These observations were puzzling until we began to examine parotid salivary gland tissues.

When experimental combinations were made between molar and parotid gland tissues, the results were not the same as with submandibular salivary gland tissues. Again, there was a very strong tendency for the glandular epithelium to express a glandular pattern of tissue branching characteristic of the parotid even when an inductively active molar papilla was used (Fig. 2.4). The lack of specificity for its own mesenchyme was seen just as Lawson (1972) and Cunha (1972) had described. Acini, tubules, and expanded epithelial cysts were present in these combinations. On the other hand

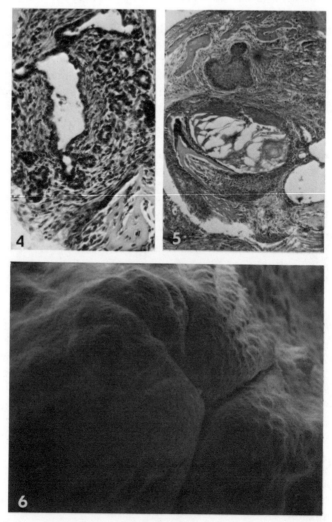

Figure 2.4 Intraocular graft of embryonic parotid epithelium and embryonic molar papilla. Maintenance of salivary gland structure is seen. (x 75)

Figure 2.5 Intraocular graft of embryonic parotid salivary gland epithelium and embryonic molar papilla. A cyst of keratinizing epithelium associated with a tooth germ is illustrated. Calcifying enamel prisms are present. (x 119)

Figure 2.6 The snout of a normal mouse embryo 16 days in utero. SEM. (x 38)

some of the epithelium formed keratinizing cysts with associated enamel organs and small tooth germs (Fig. 2.5). Thus, parotid epithelium does respond to inductive mesenchyme.

The reciprocal tissue combination of molar enamel organ and parotid mesenchyme has not resulted in an unequivocal answer to the question of inductive glandular mesenchyme. The molar epithelium produces keratinizing surface-like epithelium and, occasionally, small buds of the epithelium that are reminiscent of the early stages of salivary gland buds. But there are no clear indications of acinar branching patterns or gland-like differentiation. A range of tissues will have to be examined before a definitive conclusion can be made.

These observations suggest two things. First, that parotid and submandibular glands react differently in these experimental combinations. The parotid acts as an integumental derivative; the submandibular tissue acts as a gut derivative. These data provide experimental evidence for the notion (Hamilton and Mossman, 1972) that the parotids are of ectodermal origin and the submandibular and sublingual glands are of endodermal origin. These data, if viewed in this light, conform to the behavioral classification given in Table 2.1. The parotid should respond to the inductive stimulus of the molar and the epithelium should not be refractory to inductive stimuli. When a wider range of tissues from various ages has been tested a clearer picture of salivary gland tissue stability, responsiveness, and inductive potential will emerge. Perhaps the distinction between ectodermal and endodermal derivatives is not an absolute, and appropriate experimental conditions will be found to unlock the stability of germ layer specificity.

I repeat my comment that determination and competence may be more descriptive of investigators than embryonic tissues (Kollar, 1972).

GENETIC DEFECTS AND TISSUE INTERACTIONS

The appearance of mutations that affect the integument and oral mucosa have been examined by tissue recombination experiments in an attempt to understand the altered genetic expression, and perhaps, to gain insight into the mechanisms involved in tissue interactions. A number of mutations of skin development have been reviewed (Raphael and Pennycuik, 1980) and the localization of the defects is most often in the epithelium. However, caution should be used in interpreting these epidermal defects in light of the recent report by McAleese and Sawyer (1981) that a scaleless defect in chick skin thought to reside in the epidermis can express normal phenotype if stimulated by normal dermis of appropriate age. One exception is

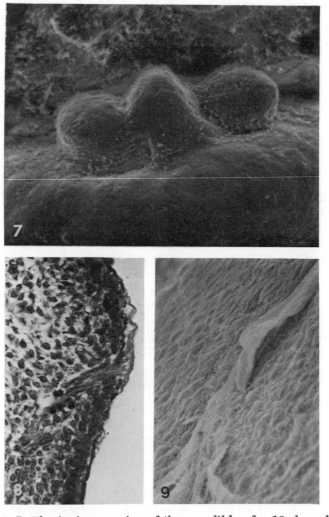

Figure 2.7 The incisor region of the mandible of a 16-day-old pf/pf mutant embryo. Note the exposed lower incisors flanking the symphysis of Meckel's cartilage. (x 38)

Figure 2.8 A section of a nerve bundle penetrating the epidermis in a pf/pf mutant embryo. (x 300)

Figure 2.9 Naked nerve fibers on the surface of the skin of a pf/pf mutant. SEM. (x 360)

hairless in which the defect is located in the dermis (Billingham and Silvers, 1968). Briggaman (1981) has investigated human genetic defects of the skin and has described the genetic defects of psoriasis and lamellar ichthyosis. These defects are most often limited to hair, vibrissae, and keratinization patterns. Expression of the genetic defects does not affect (or is not reported for) the oral mucosa or the dentition to any great extent.

On the other hand, Sofaer and Shaw (1971), Sofaer (1977a, 1977b) have described the dental defects of several mutants. These mutations most often result in altered tooth size and the presence of supernumerary teeth and appear to involve tissue interactions. Teeth seem to be very strongly conserved and are either buffered from genetic alteration, or the defects, if severe, are always lethal and therefore remain undetected in animals. The experimental usefulness of mutations that alter the integument, the oral mucosa, and the dentition is obvious.

Watson and Ede (1977) described the integumental defects associated with pupoid fetus (pf/pf), a mutant reported by Meredith (1964). In this mutant epidermal defects appear, nerves are found on the surface (Figs. 2.8 and 2.9), and epidermal proliferation attempts to heal the epidermal wounds. The repeated wound-healing results, in later stages of development, in an embryo that mimics a cocoon or pupa in shape with grossly over-developed skin nearly masking snout, limbs, and tail. The oral and dental aspects of this mutant had not been described.

A timed series of normal and mutant embryos were supplied by Dr. D. A. Ede. Figure 2.6 is an SEM view of the snout and mandible of a normal mouse. In contrast, Figure 2.7 illustrates a pf/pf littermate at the same stage of development. Wounding is very severe around the oral cavity and the mandibular incisor tooth germs and the symphysis of Meckel's cartilage are exposed. Figures 2.10 and 2.11 demonstrate the dramatic structural differences in histological sections through the snout. Note the superficial arrangement of the incisor germs and the absence of a surface epithelium (Fig. 2.11). Organized vibrissal follicles seen in the control are totally absent in the mutant. More caudally, the epithelium is present (Figs. 2.12 and 2.13) and is clearly aberrant and hyperplastic. As Watson and Ede (1977) reported, there are breaks in the epithelium associated with unusual groups of cells whose origin is uncertain. At present, it is not clear whether these are epithelial cells or papillae-like condensations of mesenchymal cells.

The oral mucosa does not demonstrate a similar histology, but there are major alterations of the tissue of the oral cavity (Figs. 2.14-2.17). The molar tooth germs are not excessively aberrant (compare Fig. 2.14 and 2.15) but there are alterations of their

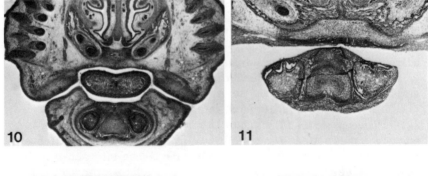

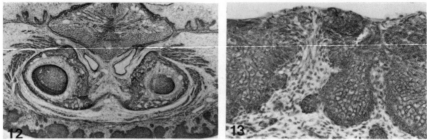

Figure 2.10 A section through the snout of a normal mouse embryo of about 18 days of gestation. The normal pattern of vibrissa, tongue, and incisor is evident. (x 17)

Figure 2.11 A section through the incisor region of a pf/pf mutant mouse embryo. The absence of a surface epithelium and the exposed incisor tooth buds are illustrated. (x 17)

Figure 2.12 A section of the oral cavity of a pf/pf mutant embryo (18-day). Note the hyperplastic surface epithelium. (x 17)

Figure 2.13 A high magnification of the surface epithelium seen in Figure 2.12. Note the perforation of the epithelium and the papilla-like aggregation of cells. (x 105)

structure—notably the absence of a dental lamina and the fusion of the upper and lower germs (compare Figs. 2.15 and 2.16). The oral cavity in general is fused. A grooved tongue was seen in every specimen as well as cleft palates (Figs. 2.15-2.16).

Needless to say, the availability of this intriguing mutant raises a number of questions. The site of expression of the mutant gene is naturally of interest. But the differences between the defect in teeth and vibrissae and the differential response of the integument and the oral mucosa raises some interesting questions about the

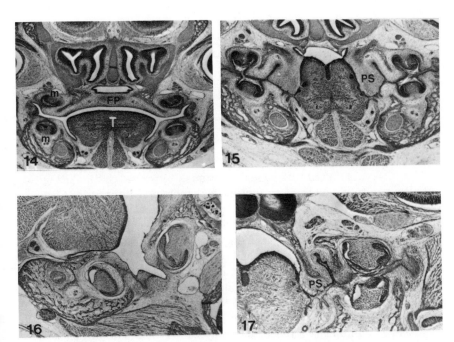

Figure 2.14 A section of a normal mouse embryo (17-day). Note the fused palate (FP), normal flattened tongue (T), and molar tooth germs (M). (x 17)

Figure 2.15 A mutant pf/pf embryo at a stage comparable to Figure 2.14. Note the cleft palate and the obstructed palatine shelves (PS), the grooved tongue, and the molar tooth germs. (x 17)

Figure 2.16 A section of the molar region of an 18-day old mouse embryo. Note the patent oral cavity (OC). (x 17)

Figure 2.17 A section of a pf/pf mutant mouse embryo comparable to Figure 2.16. Note the grooved tongue, the obstructed palatine shelves (PS) with bony elements, and the aberrant molar tooth germs. (x 17)

tissue interactions involved in the normal and mutant mouse. In addition, while the innervation seems to be involved, as suggested by preliminary examination, this seems to be a secondary phenomenon. The nerves, finding the broken epidermis, migrate to the surface. The alterations of the basement membrane that permit the migration of nerves and of mesenchymal cells onto the surface seems to be the major issue. Indeed, when this aspect of the mutation is considered, the mutation appears to mimic many of the characteristics of the human genetic disease epidermolysis bullosa which in its severest form has associated dental defects (Hintner et al., 1981).

This mutant is potentially an interesting integumental and oral deviation from normal development. The possible involvement of the basal lamina and alterations of the ECM may provide new insights into integumental and dental development.

## SUMMARY

The vertebrate integument continues to provide stimulating challenges to students of epithelial-mesenchymal interactions. A number of problems of patterning and induction of diverse integumental derivatives continue to intrigue many investigators. How instructive inductions elicit new genetic information is probably best studied in the embryonic skin and oral mucosa. What intrinsic differences exist between ectodermally and endodermally derived epithelia has not been appreciated in the context of epithelial specialization. The nature of the inductive signal is still one of the most important problems in biology. But, the role of the extracellular matrix and substrate-membrane interactions may be a reasonable alternative to diffusible inducers as a future focus of study (Kollar, 1981). The nature of the basement membrane and subjacent stroma is being described as a complex array of molecules found throughout the animal kingdom and these molecules are universally implicated in epithelial-mesenchymal interaction. How this substrate information is transduced by cells as genetic cues or as prerequisites for further interactions with the genome is an important consideration. At present it seems that mechanisms operating in the integument may be substantially and essentially different from interactions being studied in other organs. Nonetheless, the additional variations discussed here indicating that the glandular and adnexal variations in structure as well as the differences between capsular and papillar inducers may add new insights into integumental inductive events.

## ACKNOWLEDGEMENTS

I thank Dr. N. Piesco for preparing Figure 2.1 and Dr. Huw Thomas for Figures 2.2, 2.6, 2.7, and 2.9. The specimens of normal and abnormal pupoid embryos were supplied by Dr. D. A. Ede, The University of Glasgow, Scotland, to whom I am especially indebted. The University of Connecticut Research Foundation provided research funds. Mrs. S. Burke typed the manuscript with her usual pleasant efficiency.

REFERENCES

Bernfield, M. R. (1981). Organization and remodeling of the extra-
cellular matrix in morphogenesis. In Morphogenesis and Pattern
Formation (T. G. Connelly et al., eds.), pp. 138-162. Raven
Press, New York

Bernfield, M. R., Cohn, R. H., and Banerjee, S. D. (1973). Glycosaminoglycans and epithelial organ formation. Amer. Zool. 13:
1067-1083.

Billingham, R. E. and Silvers, W. K. (1968). Dermo-epidermal
interactions and epithelial specificity. In Epithelial-Mesenchymal
Interactions (R. Fleischmajer and R. E. Billingham, eds.),
pp. 252-263. Williams & Wilkins, Baltimore.

Briggaman, R. A. (1981). Basement membrane formation and origin
with special reference to skin. In Frontiers of Matrix Biology
(L. Roberts and B. Roberts, eds.), vol. 9, pp. 142-154. Karger,
Basel.

Coulombre, J. L. and Coulombre, A. J. (1971). Metaplastic induction
of scales and feathers in the corneal anterior chamber of the chick
embryo. Dev. Biol. 25:464-478.

Cunha, G. R. (1972). Support of normal salivary gland morphogenesis
by mesenchyme derived from accessory sexual glands of embryonic mice. Anat. Rec. 173:205-212.

Cunha, G. R., Lung, B., and Reese, B. (1980). Glandular epithelial
induction by embryonic mesenchyme in adult bladder epithelium of
BALB/C mice. Invest. Urol. 17:302-304.

Dhouailly, D. (1973). Dermo-epidermal interactions between birds
and mammals: differentiation of cutaneous appendages. J. Embryol. Exp. Morphol. 30:587-603.

Ekblom, P., Alitalo, K., Vaheri, A., Timple, R., and Saxen, L.
(1980). Induction of a basement membrane glycoprotein in embryonic kidney: possible role of laminin in morphogenesis. Proc.
Natl. Acad. Sci. U.S.A. 77:485-489.

Fallon, J. F. and Crosby, G. M. (1977). Polarizing zone activity in
limb buds of amniotes. In Vertebrate Limb and Somite Morphogenesis (D. A. Ede et al., eds.), pp. 55-69. Cambridge University
Press, Cambridge.

Grobstein, C. (1953). Epithelio-mesenchymal specificity in the morphogenesis of mouse submandibular rudiments in vitro. J. Exp. Zool. 124:383-414.

Grobstein, C. and Cohen, J. F. (1965). Collagenase: effect on the morphogenesis of embryonic salivary epithelium in vitro. Science 150:626-628.

Hamilton, W. J. and Mossman, H. W. (1972). Human Embryology Williams & Wilkins, Baltimore.

Hardy, M. (1968). Glandular metaplasia of hair follicles and other responses to vitamin A excess in cultures of rodent skin. J. Embryol. Exp. Morphol. 19:157-180.

Hardy, M., and Bellows, C. G. (1978). The stability of vitamin A-induced metaplasia of mouse vibrissa follicles in vitro. J. Invest. Dermatol. 71:236-241.

Hardy, M., Sweeny, P. R., and Bellows, C. G. (1978). The effects of vitamin A on the epidermis of the fetal mouse in organ culture—an ultrastructural study. J. Ultrastruc. Res. 64:246-260.

Hay, E. D. and Meier, S. (1978). Tissue interaction in development. In Textbook of Oral Biology (J. H. Shaw, E. A. Sweeny, C. C. Cappuccino, and S. M. Meller, eds.), pp. 3-23. W. B. Saunders, Philadelphia.

Hinter, H., Stingl, G., Schuler, P., Fritsch, J., Stanley, S., Katz, S., and Wolff, K. (1981). Immunofluorescence mapping of antigenic determinants within the dermal-epidermal junction in mechanobullous diseases. J. Invest. Dermatol. 76:113-118.

Hurmerinta, K., Thesleff, I., and Saxen, L. (1980). In vitro inhibition of odontoblast differentiation by vitamin A. Arch. Oral Biol. 25:385-389.

Kallenbach, E. J. and Piesco, N. P. (1978). The changing morphology of the epithelium-mesenchymal interface in the differentiation zone of growing teeth of selected vertebrates and its relationship to possible mechanisms of differentiation. J. Biol. Buccale 6: 229-240.

Kleinman, H. K., Klebe, R. J., and Martin, G. R. (1981). Role of collagenous matrices in the adhesion and growth of cells. J. Cell Biol. 88:473-485.

Kollar, E. J. (1972). The development of the integument: spatial, temporal, and phylogenetic factors. Am. Zool. 12:125-135.

Kollar, E. J. (1981). Tooth development and dental patterning. In Morphogenesis and Pattern Formation (T. G. Connelly et al., eds.), pp. 87-102. Raven Press, New York.

Kollar, E. J. and Baird, G. R. (1969). The influence of the dental papilla on the development of tooth shape in embryonic mouse tooth germs. J. Embryol. Exp. Morphol. 21:131-148.

Kollar, E. J. and Baird, G. R. (1970a). Tissue interactions in embryonic mouse tooth germs. I. Reorganization of the dental epithelium during tooth germ reconstruction. J. Embryol. Exp. Morphol. 24:159-171.

Kollar, E. J. and Baird, G. R. (1970b). Tissue interactions in embryonic mouse tooth germs. II. The inductive role of the dental papilla. J. Embryol. Exp. Morphol. 24:173-186.

Kollar, E. J. and Fisher, C. (1980). Tooth induction in chick epithelium: expression of quiescent genes for enamel synthesis. Science 207:993-995.

Kollar, E. J. and Kerley, M. (1980). Odontogenesis: interactions between isolated enamel organ epithelium and dental papilla cells. Int. J. Skelet. Res. 6:163-170.

Kollar, E. J. and Lumsden, A. S. G. (1979). Tooth morphogenesis: the role of the innervation during induction and pattern formation. J. Biol. Buccale 7:49-60.

Kratochwil, K. (1969). Organ specificity in mesenchymal induction demonstrated in the embryonic development of the mammary gland of the mouse. Dev. Biol. 20:46-71.

Lawson, K. A. (1972). The role of mesenchyme in morphogenesis and functional differentiation of the salivary epithelium. J. Embryol. Exp. Morphol. 27:497-513.

Lesot, H., van der Mark, V., and Ruch, J. V. (1978). Localization par immunofluorescence des types de collagene synthesis par l'ebauche dentaire chez l'embryon de souris. C. R. Acad. Sci. 286:765-768.

McAleese, S. R. and Sawyer, R. H. (1981). Correcting the phenotype of the epidermis from chick embryos homozygous for the gene
scaleless (sc/sc). Science 214:1033-1034.

Meredith, R. (1964). Research news. The Mouse Newsletter 31:25.

Raphael, K. A. and Pennycuik, P. R. (1980). The site of action of
the naked locus (N) in the mouse as determined by dermal-
epidermal recombinations. J. Embryol. Exp. Morphol. 57:143-
153.

Rawles, M. E. (1963). Tissue interactions in scale and feather development as studied in dermal-epidermal recombinations. J. Embryol. Exp. Morphol. 11:765-789.

Sakakura, T., Nishizuka, Y., and Dawe, C. J. (1976). Mesenchyme-
dependent morphogenesis and epithelium-specific cytodifferentiation in mouse mammary gland. Science 194:1439-1441.

Saxen, L., Ekblom, P., and Lehtonen, E. (1981). The kidney as a
model system for determination and differentiation. In The Biology
of Normal Human Growth (M. Ritzen et al., eds.), pp. 117-127.
Raven Press, New York.

Saxen, L., Ekblom, P., and Thesleff, I. (1980). Mechanisms of
morphogenetic cell interactions. In Development in Mammals,
Vol. 4 (M. H. Johnson, ed.), pp. 161-201. Elsevier/North-
Holland, New York.

Sengel, P. (1976). In Morphogenesis of Skin, pp. 1-277. Cambridge
University Press, Cambridge.

Slavkin, H. C. (1978). The nature and nurture of epithelio-mesenchymal interactions during tooth morphogenesis. J. Biol. Buccale
6:189-204.

Slavkin, H. C., Yamada, M., Bringas, P., and Grodin, M. (1980).
Tooth epithelial differentiation in vitro and congenital craniofacial
malformation. In Birth Defects: Original Article Series, Volume
XVI, Number 2, pp. 211-230. Alan R. Liss, Inc., New York.

Sofaer, J. A. (1977a). The teeth of the 'sleek' mouse. Arch. Oral
Biol. 22:299-301.

Sofaer, J. A. (1977b). Tooth development in the 'crooked' mouse. J. Embryol. Exp. Morphol. 41:279-287.

Sofaer, J. A. and Shaw, J. H. (1971). The genetics and development of fused and supernumerary molars in the rice rat. J. Embryol. Exp. Morphol. 26:99-109.

Thesleff, I. and Hurmerinta, K. (1981). Tissue interactions in tooth development. Differentiation 18:75-88.

Thesleff, I., Lehtonen, E., and Saxen, L. (1978). Basement membrane formation in transfilter tooth culture and its relationship to odontoblast differentiation. Differentiation 10:71-79.

Watson, P. J. and Ede, D. A. (1977). Neural-epidermal interactions in the developing skin of the pupoid-fetus (pf/pf) mouse mutant embryo. J. Anat. 124:229.

Wolpert, L. (1981). Positional information, pattern formation, and morphogenesis. In Morphogenesis and Pattern Formation (T. G. Connelley et al., eds.), pp. 5-20. Raven Press, New York.

# 3

# EPITHELIAL-MESENCHYMAL INTERACTIONS IN HORMONE-INDUCED DEVELOPMENT

*Gerald R. Cunha, John M. Shannon,*
*Osamu Taguchi, Hirohiko Fujii,*
*and Beth A. Meloy*

Morphogenetic processes within the integumental, gastro-
intestinal, and urogenital systems are strictly dependent upon interactions between epithelium and mesenchyme (Kollar, 1972; Sengel,
1976; Yasugi and Mizuno, 1978; Mizuno et al., 1975; Cunha, 1976a;
Cunha et al., 1980a; 1981). For all of these systems mesenchyme
induces and specifies epithelial morphogenesis and cytodifferentiation.
While urogenital morphogenesis is regulated by sex steroids and development of most other organs is not, the basic mechanism mediating
these epithelial-mesenchymal interactions, though poorly understood,
is probably similar. For instance, epithelium of the salivary gland,
whose early development is independent of sex hormones, exhibits
acinar development and normal cytodifferentiation when combined
with mesenchyme from either the developing salivary gland, prostate,
or seminal vesicle (Lawson, 1972; Cunha, 1972c; Ball, 1974). These
findings indicate that both salivary and urogenital mesenchymes provide similar physicochemical conditions that are permissive for
salivary epithelial development. Instructive inductions, in which epithelial development is reprogrammed by a foreign (heterotypic)
mesenchyme, also appear to be mediated by similar developmental
mechanisms in hormore-target versus nontarget tissues. For instance, the inductive signals from urogenital sinus mesenchyme
elicit prostatic morphogenesis and cytodifferentiation in epithelium
from either the urogenital sinus (a known hormone target) or the
urinary bladder, whose development is independent of hormones
(Cunha and Lung, 1978; Cunha et al., 1980a; 1981). Thus, although
the fundamental mechanisms mediating epithelial development are
probably similar in hormone-target versus nontarget organs, the

expression of hormonal sensitivity in developing urogenital organs remains to be explained.

In adulthood, epithelia of accessory sexual glands constitute the secretory parenchyma of the genital tract. Secretory activity within urogenital epithelia is elicited by the appropriate sex hormone(s) whose action is mediated by specific, high affinity hormone receptor proteins (Clark and Peck, 1979; Mainwaring, 1977). Sex steroids enter secretory epithelial cells by diffusion and bind to receptor proteins within the cytoplasm. The resultant hormone-receptor complex then translocates to the nucleus, and following interaction with acceptor sites on the chromatin, transcription of mRNA's coding for secretory products is stimulated, which ultimately leads to their synthesis and release (O'Malley and Schrader, 1976; Higgins et al., 1976; Higgins and Parker, 1980; Parker et al., 1978; Mainwaring, 1977). By contrast, during embryonic and neonatal periods the remarkable secretory effects elicited by sex hormones in adult epithelial cells are noticeably absent following hormonal stimulation. Instead, epithelia of developing urogenital rudiments are involved in morphogenetic processes that ultimately lead to the differentiation of a secretory epithelium. Insofar as epithelial morphogenesis is induced and specified by mesenchymal cells, urogenital epithelia can be ascribed a relatively passive role during developmental periods (Cunha, 1976a; Cunha and Lung, 1979; Cunha et al., 1980a; 1981). Given the facts that hormones affect morphogenesis of urogenital glands and that mesenchyme is the inductor of epithelial development, it is reasonable to ascribe the hormonal sensitivity of developing urogenital rudiments to the mesenchyme. Support for this concept is illustrated below.

Tissue recombination studies on the developing prostate and mammary gland utilizing epithelium and mesenchyme from androgen-sensitive wild-type and androgen-insensitive Tfm mice (Testicular feminization) directly implicate the mesenchyme as the target and mediator of androgenic effects upon epithelial development. Male Tfm mice do not exhibit masculine morphogenesis because of an insensitivity to androgens based on an absence, reduction, or defect in androgen receptor activity (Bardin and Bullock, 1974; Ohno, 1977; Wilson et al., 1981). When embryonic wild-type epithelium is grown in association with Tfm mesenchyme, androgen-induced morphogenesis does not occur. By contrast, when Tfm epithelium is grown in association with wild-type mesenchyme, androgen-induced morphogenesis is expressed (Fig. 3.1). These experiments demonstrate that it is the mesenchyme that is the actual target and mediator of androgenic effects upon the epithelium (Kratochwil and Schwartz, 1976; Drews and Drews, 1977; Cunha and Lung, 1978; Lasnitzki and Mizuno, 1980). This finding corroborates earlier tissue recombination

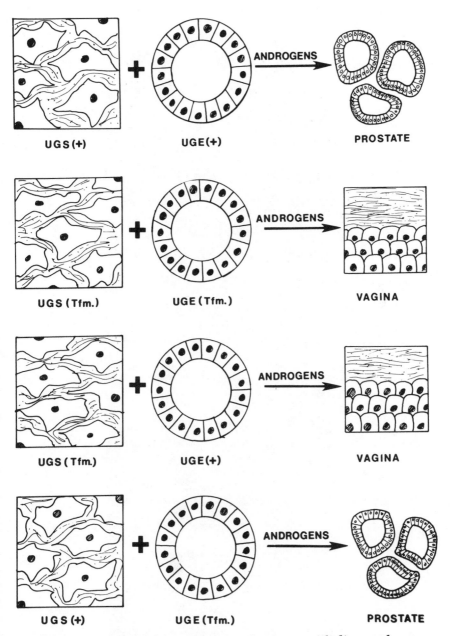

UGS(+)   UGE(+)   ANDROGENS   PROSTATE

UGS(Tfm.)   UGE(Tfm.)   ANDROGENS   VAGINA

UGS(Tfm.)   UGE(+)   ANDROGENS   VAGINA

UGS(+)   UGE(Tfm.)   ANDROGENS   PROSTATE

Figure 3.1 Analysis of recombinants between epithelium and mesenchyme of urogenital sinuses from androgen-insensitive Tfm and
androgen-responsive wild-type mouse embryos. Exposure to androgens was accomplished by grafting the recombinants to male hosts.
Note that the androgenic response (prostatic morphogenesis) occurs
only in those recombinants constructed with wild-type mesenchyme.
By contrast, when Tfm mesenchyme is used, androgen-induced
morphogenesis is absent, and the epithelium exhibits vaginal differentiation. These observations indicate that the mesenchyme is the
actual target and mediator of androgenic effects upon the epithelium
(Cunha et al., 1980a).

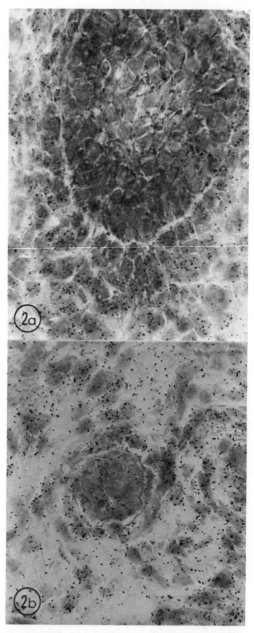

Figure 3.2a and 3.2b  Autoradiograms of $^3$H-dihydrotestosterone
localization in: (a) the 14-day embryonic female mouse mammary
gland; and (b) the 4-day neonatal mouse prostate. Note that nuclear
labeling, indicative of androgen receptor activity, is prominent in
mesenchymal cells, but absent within the epithelium (a, x 1280;
b, x 960) (Cunha et al., 1981).

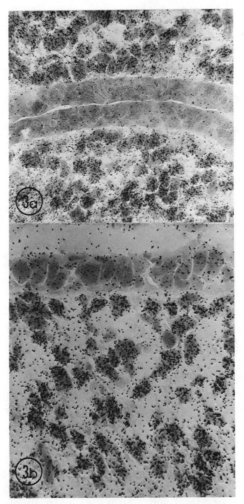

Figure 3.3a and 3.3b Autoradiograms of $^3$H-estradiol localization in: (a) the uterus; and (b) sinus vagina of 4 day old mice. Nuclear labeling (estrogen-receptor activity) is prominent within mesenchymal cells but absent in epithelial cells. (a, x 800; b, x 1280)

studies between urogenital and integumental tissues which also suggest that urogenital mesenchyme is the mediator of androgenic effects upon epithelial development (Cunha, 1970; Cunha, 1972a; 1972b; 1972c). Since androgenic effects are mediated by androgen receptor proteins, the conclusion that mesenchyme is the target and mediator of androgenic effects upon the epithelium implies that expression of androgen receptor activity within mesenchymal cells is obligatory for

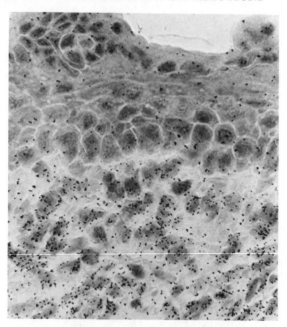

Figure 3.4 An autoradiogram of $^3$H-estradiol in the lower (sinus) vagina of a neonatal mouse that received 4 daily injections of diethylstilbestrol (DES) (5 ug/day) from birth and was sacrificed 24 hours after the last injection. DES treatment has elicited vaginal epithelial proliferation and cornification, which is expressed in the epithelium despite the continued absence of estrogen receptors (nuclear labeling). Note that the mesenchymal cells exhibit estrogen-receptor activity. (x 1280)

the development of androgen target organs. Recent autoradiographic analysis of nuclear androgen binding in the developing prostate and mammary gland of wild-type mice indicates that androgen receptor activity is present only within the mesenchymal cells of these developing organ rudiments (Fig. 3.2a and 3.2b) (Shannon et al., 1981; Cunha et al., 1981). In fact, epithelial cells of the developing mammary gland and prostate (urogenital sinus) lack androgen receptor activity, suggesting that the expression of androgen-induced epithelial morphogenesis is regulated by nonsteroidal inductors elaborated by mesenchymal cells following androgenic stimulation. The expression of androgen-induced prostatic morphogenesis and function in androgen-insensitive Tfm epithelium also supports this concept (Cunha and Chung, 1982). We, therefore, believe that expression of hormonal sensitivity in morphogenetic processes is a consequence of the acquisition of hormone-receptor activity within mesenchyme, which is

the inductor of epithelial morphogenesis in all systems examined.
Finally, if hormone-sensitivity in epithelial development is a function
of expression of hormone receptor activity within mesenchyme, sensitivity to estrogenic hormones may also be regulated in a similar
fashion. Indeed, in the developing uterus, cervix, and vagina (organs
whose morphogenesis is profoundly influenced by exogenous estrogens
(McLachlan et al., 1981; Forsberg and Kalland, 1981; Takasugi,
1976; Bern and Talamantes, 1981), autoradiographic evidence indicates that estrogen-receptor activity is prominently displayed in
mesenchymal cells (Fig. 3.3a and 3.3b), while estrogen receptor
activity is undetectable in the epithelium (Stumpf et al., 1980; Cunha
et al., 1982b). In fact, vaginal cornification is elicited in the sinus
vagina* of neonatal mice by exogenous estrogen within an epithelium
that continues to remain devoid of estrogen receptor activity (Fig.
3.4) (Cunha et al., 1982a). Thus, analysis of developing androgen
and estrogen target organs indicates that hormonal dependence of
morphogenetic processes is related to the expression of hormone-
receptor activity within mesenchymal cells, the inductors of epithelial development.

Given the facts that mesenchyme induces epithelial morphogenesis and is responsible for the expression of hormonal sensitivity
during urogenital development, the epithelial response to mesenchymal induction will now be explored. Epithelial morphogenesis and
histogenesis is induced and specified within the urogenital tract by
the mesenchyme (Cunha, 1976a). For instance, differentiation of
embryonic Mullerian epithelium into a simple columnar, uterine
epithelium or into a stratified squamous, potentially cornified vaginal
or cervical epithelium (Fig. 3.5) is induced by mesenchyme of the
neonatal uterus, vagina, and cervix, respectively (Cunha, 1976b;
Cunha et al., 1980a; 1981). The profound histological and cytological
changes elicited within Mullerian epithelium by heterotypic urogenital
stromas suggest the possibility that these morphological changes may
be coupled with functional changes in the epithelium of the female
genital tract. This possibility has recently been explored through
analysis of tissue recombinants prepared with epithelial and mesenchymal components of the neonatal uterus and vagina. Uterine mesenchyme (UM) elicits uterine morphogenesis when combined with either
uterine epithelium (UE) or vaginal epithelium (VE). Similarly, vaginal
mesenchyme (VM) elicits vaginal differentiation in either VE or UE
(Cunha, 1976b). Analysis of two-dimensional polyacrylamide gels of
radio-labeled proteins in adult uterus and vagina (as intact organs or

---

*The sinus vagina in mice constitutes the lower two-fifths of
this organ and is derived from the urogenital sinus.

separated tissues) demonstrates that each organ exhibits a unique, characteristic protein synthetic "finger print" that is primarily accounted for by the biosynthetic activity of the epithelial component (data not shown). Comparison of gels of uterus and vagina with those of heterotypic recombinants (UM + VE → Uterus; and VM + UE → Vagina) indicates that protein synthetic activity corresponds to the induced epithelial phenotype. Gels of UM + VE recombinants resemble gels of uterus, and gels of VM + UE recombinants resemble gels of

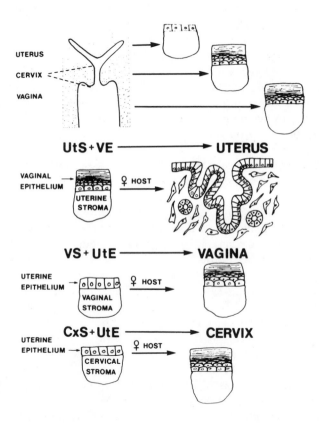

Figure 3.5 Summary of heterotypic recombinations between epithelium and mesenchyme of the uterus, cervix, and vagina of neonatal mice. Glandular (uterine) differentiation of undifferentiated Mullerian epithelium of the developing vagina is elicited by recombination with uterine mesenchyme. Squamous differentiation is elicited in uterine epithelium by either vaginal or cervical mesenchyme. Thus, the regional differentiation of Mullerian epithelium into uterus, cervix, and vagina is specified by the mesenchyme (Cunha and Fujii, 1981).

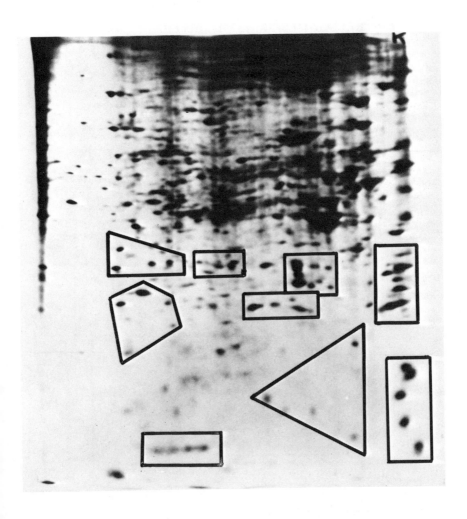

Figure 3.6a, 3.6b, 3.6c, and 3.6d Two-dimensional polyacrylamide gel electrophoresis of $^{35}$S-methionine labeled proteins in: (a) adult uterus; (b) UM + VE recombinants; (c) adult vagina; and (d) VS + UE recombinants. Comparison of gels of adult uterus (a) with adult vagina (c) loaded with comparable amounts of protein reveals that each organ exhibits a distinctive protein synthetic "fingerprint." Note that the gel of the UM + VE recombinant (b), which yields uterine epithelial differentiation, displays a protein synthetic "fingerprint" that corresponds to that of uterus (a), while the VS + UE gel (d) resembles that of vagina (c). These data suggest that mesenchyme-induced alterations in epithelial morphogenesis and cytodifferentiation are coupled with predictable changes in functional (biosynthetic) activity.

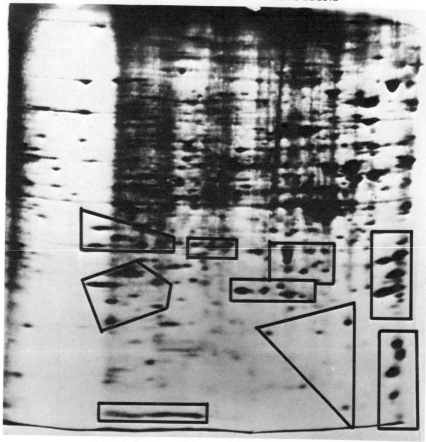

Figure 3.6b  See legend on page 59.

vagina (Fig. 3.6a, b, c, d). We, therefore, suggest that epithelial cytodifferentiation is coupled to changes in functional (biochemical) activity.

Further support for this idea is provided through analysis of heterotypic recombinants composed of urogenital sinus mesenchyme (UGM) and bladder epithelium (BLE) (UGM + BLE recombinants) which, when grown in male hosts, give rise to prostatic tissue (Cunha and Lung, 1978; Cunha et al., 1980a; 1981). Histological and fine structural evidence of secretory activity within the induced prostatic acini of UGM + BLE recombinants suggests that mesenchyme-induced alteration in epithelial histodifferentiation may be coupled to alterations in functional (biochemical) activity. Histochemical analysis demonstrates that markers indicative of urothelium (bladder epithe-

lium) are lost concomitant with the expression of prostatic markers
in acini induced in UGM + BLE recombinants (Table 3.1) (Cunha and
Lung, 1980). Moreover, attainment of prostatic secretory differentiation in urothelium is associated with the expression of androgen
receptor activity in the induced epithelium (Cunha et al., 1980c).
Androgen receptor activity, which is a distinctive feature of prostatic differentiation, has been demonstrated within epithelium of
UGM + BLE recombinants by autoradiographic (Cunha et al., 1980c)
and biochemical studies (Thompson et al., in preparation). Expression of androgen receptor activity within urothelium appears to be
the basis for androgen-inducibility of DNA synthesis in UGM + BLE

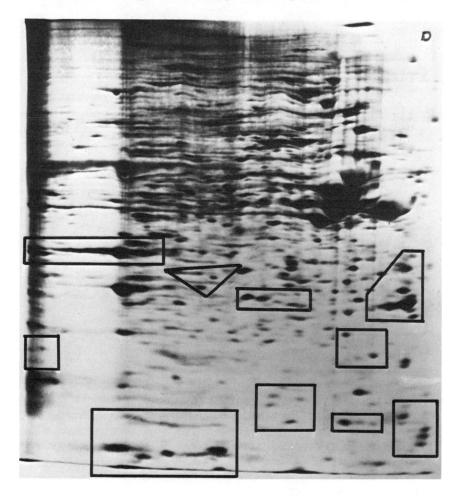

Figure 3.6c  See legend on page 59.

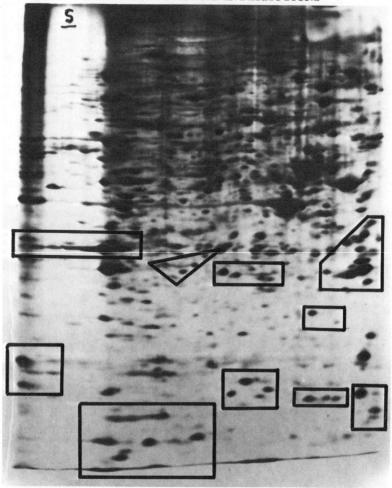

Figure 3.6d See legend on page 59.

recombinants, which predictably is blocked by coadministration of
the antiandrogen, cyproterone acetate (Cunha and Chung, 1982).
Since cyproterone acetate blocks androgenic effects by competition
of androgen binding at the site of the androgen receptor (Fang and
Liao, 1969), this provides additional evidence for the expression of
androgen-receptor activity in the prostatic acini induced in urothelium
by urogenital sinus mesenchyme. Investigation of protein synthesis
in UGM + BLE recombinants by two-dimensional gel electrophoresis
also suggests that a true (biochemical) prostatic induction has occurred. Although analysis of protein synthesis in homogenates of

TABLE 3.1

Histochemical Analysis of Prostate, Bladder, and Tissue
Recombinants Prepared with UGM[a] and BLE[b]
of Embryonic Mice

| Specimen | Epithelial Morphology | Histochemical characteristics of epithelia | | |
|---|---|---|---|---|
| | | Alcian Blue | Nonspecific Esterase | Alkaline Phosphatase |
| Adult prostate | Glandular | + | + | − |
| Adult urinary bladder | Transitional | − | − or ± | + |
| UGM + BLE | Glandular | + | + | − |

[a]UGM—urogenital sinus mesenchyme.
[b]BLE —epithelium from urinary bladder.
Source: Cunha and Lung (1980).

prostate, bladder, and UGM + BLE recombinants cannot specify the
tissue (epithelium or stroma) responsible for the synthetic activity,
UGM + BLE recombinants exhibit protein synthetic profiles that are
distinctly prostatic (Cunha and Chung, 1982). Finally, immunocytochemical analysis of UGM + BLE recombinants confirms the expression of prostate-specific antigens in the induced bladder epithelium
(Cunha et al., 1981; Taguchi and Cunha, in preparation). We, therefore, conclude that the inductive influences of urogenital sinus mesenchyme reprogram both the morphological as well as the functional
(biochemical) characteristics of bladder epithelium. Thus, mesenchyme-induced changes in epithelial cytodifferentiation in male and
probably female genital tracts are coupled to changes in functional
activity.

Similar findings have been reported for lens and scale induction.
Differentiation of lens fiber cells from epidermis grown in association
with optic cup results in the expression of lens crystallins in the
induced epithelial cells (Karkinen–Jaaskelainen, 1978). Similarly,
mesenchyme-induced formation of scales from feather epidermis
results in the expression of scale-specific keratins (Dhouailly et al.,
1978). The apparent coupling of cytodifferentiation and functional
(biochemical) differentiation in the examples cited above appear to be

in direct contrast to Sakakura's observations on mammary development (Sakakura et al., 1976). These investigators have demonstrated that salivary gland mesenchyme elicits a lobular branching pattern in mammary epithelium that is characteristic of salivary gland. Such structures, which at a gross level resemble salivary gland, were shown to be able to produce a milk protein. The apparent contradiction between the studies on lens, scale, prostatic, uterine, and vaginal inductions and that reported for the mammary gland may be explained by the fact that in the former cases a true change in epithelial cytodifferentiation occurred (e.g., transitional urothelial cells became a simple columnar glandular epithelium) while in the case of the salivary-mammary recombinants there is no evidence that epithelial cytodifferentiation was altered. Instead, only the gross lobular branching pattern of mammary epithelium was modified to a pattern resembling salivary gland. Thus, in heterotypic tissue recombinations mesenchyme induces and may reprogram epithelial morphogenesis and cytodifferentiation, which in turn may lead to the expression of new patterns of functional (biochemical) activity.

Although the mechanism of epithelial-mesenchymal interactions is unknown, tissue recombination studies between epithelium and mesenchyme from different species (rat, mouse, rabbit, and human) demonstrate that the mechanisms of urogenital development are similar, if not identical, in all of these species (Table 3.2). Recombination of urogenital sinus mesenchyme and epithelium (permissive

TABLE 3.2

Heterospecific Induction of Prostatic or Glandular Differentiation by Urogenital Sinus Mesenchyme

| Specimen | | Result |
|---|---|---|
| Mesenchyme | Epithelium | |
| Rat UGM | Mouse UGE | Prostate (15/15) |
| Mouse UGM | Rat UGE | Prostate (3/3) |
| Rat UGM | Rabbit UGE | Glands (6/6) |
| Mouse UGM | Rabbit UGE | Glands (8/8) |
| Rabbit UGM | Mouse UGE | Prostate (10/10) |
| Rat UGM | Mouse BLE | Prostate (15/15) |
| Mouse UGM | Rabbit BLE | Prostate (2/2) |
| Mouse UGM | Rat BLE | Prostate (7/7) |
| Mouse UGM | Human fetal BLE | Developing prostate? (40/40) |

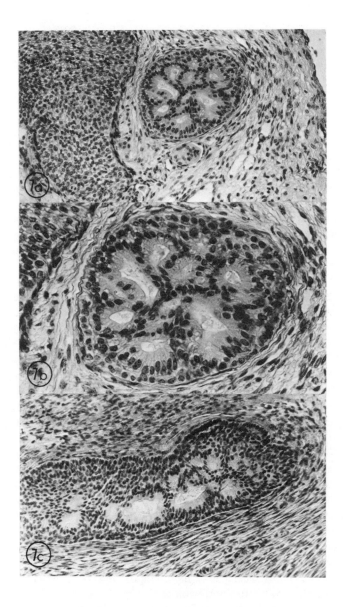

Figure 3.7a  A recombinant composed of mouse urogenital sinus mesenchyme and human fetal bladder epithelium which was grown in a male athymic nude mouse for four months. The human fetal urothelium has been induced to form glandular structures resembling embryonic human prostatic tubules. (x 240)

Figure 3.7b  Higher magnification of (a). Note that the induced human urothelial cells are differentiating into a tall columnar secretory epithelium. (x 480)

Figure 3.7c  Human fetal prostate from a 10 week old fetus grown for 1 month in a male athymic nude mouse. Note that the solid prostatic cords are undergoing canalization and that the luminal epithelial cells are differentiating into tall columnar secretory epithelial cells (x 320) (Cunha et al., 1981).

induction) from different species leads to normal prostatic development (Cunha, unpublished). Similarly, in heterotypic (instructive) inductions between urogenital sinus mesenchyme and bladder epithelium, prostatic differentiation is induced across species lines. The demonstration that mouse urogenital sinus mesenchyme can elicit glandular (prostatic?) differentiation (Fig. 3.7a, b, c) from human fetal bladder epithelium suggests that developmental mechanisms in humans and laboratory animals are similar, if not identical. This emphasizes the validity of animal models as tools for understanding human birth defects.

Though epithelial-mesenchymal interactions are primarily associated with embryonic development, most (if not all) of the concepts described above are also applicable in adulthood. Regional patterns of adult epidermal differentiation are controlled by dermal cells (Billingham and Silvers, 1968; Briggaman and Wheeler, 1971; Oliver, 1968; Karring et al., 1975; Friedenwald, 1951; Bernimoulin and Schroeder, 1977). In hormone-target organs such as the mammary gland (Sakakura et al., 1979a; Daniel and De Ome, 1965) and vagina (Cunha, 1976b; Cunha et al., 1981), adult epithelium is capable of participating in mesenchyme-induced morphogenetic processes. Adult stromal cells from the vagina, mammary gland, and prostate can function as inductors thereby eliciting epithelial morphogenesis (Sakakura et al., 1979a) and/or reprogramming epithelial differentiation (Cunha, 1976b; Cunha et al., 1981). Moreover, epithelia of the adult urinary bladder and vagina have been shown to be responsive to prostatic induction by urogenital sinus mesenchyme, which elicits a true prostatic epithelial differentiation from urothelium as judged by histological, histochemical, fine structural, immunocytochemical, steroid autoradiographic and biochemical analyses (Cunha et al., 1980a, 1980b; 1981; 1982a; Cunha and Fujii, 1981; Cunha and Chung, 1982). Therefore, in adulthood many aspects of epithelial morphology and functional activity are under stromal control by mechanisms that are probably similar to those that are operative during embryonic development. This finding may have important implications for abnormal epithelial differentiation and growth.

A relationship between epithelial-stromal interactions and the development of epithelial neoplasms has been proposed by several investigators. One theory suggests that carcinogenic agents may elicit primary, non-neoplastic effects within the stroma, which in turn mediate neoplastic development in the epithelium. Support for this hypothesis comes from Hodges et al. (1977) who demonstrated that "preneoplastic" cell surface architecture is expressed in normal bladder epithelial cells when grown in association with stroma from carcinogen-treated bladders. The reversion to a normal epidermis of epithalia from basal cell carcinomas (Cooper and Pinkus, 1977) and

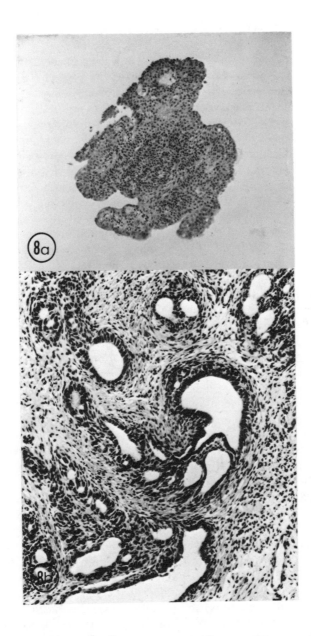

Figure 3.8a  A bladder papilloma induced in a rat with the carcinogen N-methyl-nitrosourea. The papilloma is covered with undifferentiated neoplastic epithelial cells diagnosed as a nonglandular transitional cell carcinoma. Note the absence of glandular elements in the tumor. (x 200)

Figure 3.9b  A tissue recombinant composed of embryonic urogenital sinus mesenchyme and epithelium from a transitional cell carcinoma as depicted in (a). Growth of the recombinant in a male host has resulted in the induction of adenocarcinomatous differentiation in the nonglandular transitional carcinoma cells. (x 400)

leukoplakia (MacKenzie et al., 1979) following association with a normal connective tissue further emphasizes the role of stroma as a mediator of the carcinogenic process. Another theory suggests that neoplastic change occurs primarily within the epithelium, but that the stroma (mesenchyme) is involved in determining and controlling the expression of the biological character of the neoplasm. This idea is supported by the studies of Dawe and his colleagues on polyoma-induced transformation of the mouse submandibular gland (Dawe, 1971; Dawe et al., 1976). From studies on mammary carcinogenesis, Sakakura et al. (1979b; 1981) have further demonstrated that mesenchyme can accelerate tumor development without altering the biological and morphological character of the resultant neoplasms. Our recent studies on tumors of the urinary bladder now indicate that major morphologic alterations can be induced within epithelial neoplasms by heterotypic mesenchyme without reversion of the epithelium to a normal phenotype (Fujii et al., 1982). For example, transitional cell carcinomas induced in rats with N-methyl-nitrosourea exhibit a nonglandular bladder phenotype in situ or in grafts composed of the neoplastic epithelium and its associated stroma (Fig. 3.8a). By contrast, when neoplastic urothelium from these transitional cell carcinomas is grown in association with urogenital sinus mesenchyme (a prostatic inductor), adenocarcinomatous differentiation is elicited in the normally nonglandular epithelial cells (Fig. 3.8b) (Fujii et al., 1982). This change in epithelial cytodifferentiation does not appear to be accompanied by a reversion to a state of normalcy. On the contrary, the altered epithelial cells continue to exhibit several features indicative of neoplasia. However, since changes in epithelial cytodifferentiation appear to be coupled with changes in functional activity (see earlier discussion on UGM + BLE recombinants), we speculate that the biological activity and potential of the induced adenocarcinomatous structures may be correspondingly altered.

In conclusion, regulation of epithelial development and differentiation by mesenchymal cells is a common theme utilized during the development of a variety of organ systems. In the urogenital tract, whose development is hormone-dependent, this is accomplished through the inductive activities of mesenchymal cells that exhibit androgen and/or estrogen receptor activity. For adult hormone-target organs, which express marked cyclical changes in epithelial morphogenesis, cytodifferentiation and functional activity during estrous, menstrual, or reproductive cycles, epithelial-mesenchymal interactions play important morphogenetic roles in adulthood. Perturbation of these tissue interactions may influence carcinogenic processes.

ACKNOWLEDGEMENTS

The authors are indebted to Ken D. Vanderslice, Marnie Sekkingstad, Caroline Montoya, and Virginia Miller for their technical assistance and to Judy Paden and Betty Aguilar for typing the manuscript. This study was supported by the following grants: ACS Grant #PDT-139; March of Dimes Grant #1-670; NIH Grants #HD12116, AM25266, CA27418, AMCA16570.

REFERENCES

Ball, W. D. (1974). Development of the rat salivary glands. III. Mesenchymal specificity in the morphogenesis of the embryonic submaxillary and sublingual glands of the rat. J. Exp. Zool. 188:277-288.

Bardin, C. W. and Bullock, L. P. (1974). Testicular feminization: Studies of the molecular basis of a genetic defect. J. Invest. Dermatol. 63:75-84.

Bern, H. A. and Talamantes, F. J. (1981). Neonatal mouse models and their relation to disease in the human female. In Developmental Effects of Diethylstilbestrol (DES) in Pregnancy (A. L. Herbst and H. A. Bern, eds.), pp. 129-147. Thieme-Stratton, New York.

Bernimoulin, J.-P. and Schroeder, H. E. (1977). Changes of differentiation pattern in oral mucosal epithelium following heterotypic connective tissue transplantation in man. Pathol. Res. Pract. 166:290-312.

Billingham, R. E. and Silvers, W. K. (1968). Dermoepidermal interactions and epithelial specificity. In Epithelial-Mesenchymal Interactions (R. Fleischmajer and R. E. Billingham, eds.), pp. 252-266. Williams & Wilkins, Baltimore.

Briggaman, R. A. and Wheeler, C. E. (1971). Epidermal-dermal interactions in adult human skin. II. The nature of the dermal influence. J. Invest. Dermatol. 56:18-26.

Clark, J. H. and Peck, E. J., Sr. (1979). Female sex steroids: Receptors and function. Monogr. Endocrinol., Vol. 14. Springer Verlag, New York.

Cooper, M. and Pinkus, H. (1977). Intrauterine transplantation of rat basal cell carcinoma as a model for reconversion of malignant to benign growth. Cancer Res. 37:2544-2552.

Cunha, G. R. (1970). Epithelio-mesenchymal interactions in the developing accessory sexual glands of the mouse embryo. Anat. Rec. 166:295 (Abstract).

Cunha, G. R. (1972a). Tissue interactions between epithelium and mesenchyme of urogenital and integumental origin. Anat. Rec. 172:529-542.

Cunha, G. R. (1972b). Epithelio-mesenchymal interactions in primordial gland structures which become responsive to androgenic stimulation. Anat. Rec. 172:179-196.

Cunha, G. R. (1972c). Support of normal salivary gland morphogenesis by mesenchyme derived from accessory sexual glands of embryonic mice. Anat. Rec. 173:205-212.

Cunha, G. R. (1976a). Epithelial-stromal interactions in development of the urogenital tract. Int. Rev. Cytol. 47:137-194.

Cunha, G. R. (1976b). Stromal induction and specification of morphogenesis and cytodifferentiation of the epithelia of the Mullerian ducts and urogenital sinus during development of the uterus and vagina in mice. J. Exp. Zool. 196:361-370.

Cunha, G. R. and Chung, L. W. K. (1982). Stromal-epithelial interactions: Induction of prostatic phenotype in urothelium of testicular feminized (Tfm) mice. J. Steroid Biochem. (in press).

Cunha, G. R. and Fujii, H. (1981). Stromal-parenchymal interactions in normal and abnormal development of the genital tract. In Developmental Effects of Diethylstilbestrol (DES) in Pregnancy (A. L. Herbst and H. A. Bern, eds.), pp. 179-193. Thieme-Stratton, New York.

Cunha, G. R. and Lung, B. (1978). The possible influences of temporal factors in androgenic responsiveness of urogenital tissue recombinants from wild-type and androgen-insensitive (Tfm) mice. J. Exp. Zool. 205:181-194.

Cunha, G. R. and Lung, B. (1979). The importance of stroma in morphogenesis and functional activity of urogenital epithelium. In Vitro 15:50-71.

Cunha, G. R. and Lung, B. (1980). Experimental analysis of male accessory sex gland development. In Accessory Glands of the Male Reproductive Tract (E. Spring-Mills and E. S. E. Hafez, eds.), pp. 39-59. Elsevier/North-Holland, New York.

Cunha, G. R., Chung, L. W. K., Shannon, J. M., and Reese, B. A. (1980a). Stromal-epithelial interactions in sex differentiation. Biol. Reprod. 22:19-42.

Cunha, G. R., Lung, B., and Reese, B. (1980b). Glandular epithelial induction by embryonic mesenchyme in adult bladder epithelium of BALB/c mice. Invest. Urol. 17:302-304.

Cunha, G. R., Reese, B. A., and Sekkingstad, M. (1980c). Induction of nuclear androgen-binding sites in epithelium of the embryonic urinary bladder by mesenchyme of the urogenital sinus of embryonic mice. Endocrinology 107:1767-1770.

Cunha, G. R., Shannon, J. M. Neubauer, B. L., Sawyer, L. M., Fujii, H., Taguchi, O., and Chung, L. W. K. (1981). Mesenchymal-epithelial interactions in sex differentiation. Hum. Genet. 56:68-77.

Cunha, G. R., Shannon, J. M., Taguchi, O  Fujii, H., and Chung, L. W. K. (1982a). Mesenchymal-epithelial interactions in hormone-induced development. J. Animal Science (in press).

Cunha, G. R., Shannon, J. M., Vanderslice, K. D., Sekkingstad, M., and Robboy, S. J. (1982b). Autoradiographic analysis of nuclear estrogen binding sites during postnatal development of the genital tract of female mice. J. Steroid Biochem. (submitted).

Daniel, C. W. and De Ome, K. B. (1965). Growth of mouse mammary glands in vivo after monolayer culture. Science 149:634-636.

Dawe, C. J. (1971). Epithelial-mesenchymal interactions in relation to the genesis of polyoma virus-induced tumors of mouse salivary gland. In Tissue Interactions in Carcinogenesis (D. Tarin, ed.), pp. 305-358. Academic Press, New York.

Dawe, C. J., Morgan, W. D., Williams, J. E., and Summerour, J. P. (1976). Inductive epithelio-mesenchymal interaction: Is it involved in the development of epithelial neoplasms? In Progress in Differentiation Research (N. Muller-Berat et al., eds.), pp. 305-318. Elsevier/North-Holland, Amsterdam.

Dhouailly, D., Rogers, G. E., and Sengel, P. (1978). The specification of feather and scale protein synthesis in epidermal-dermal recombinations. Dev. Biol. 65:58-68.

Drews, U. and Drews, U. (1977). Regression of mouse mammary gland anlagen in recombinants of Tfm and wild-type tissues: Testosterone acts via the mesenchyme. Cell 10:401-404.

Fang, S. and Liao, S. (1969). Antagonistic action of antiandrogens on the formation of a specific dihydrotestosterone receptor complex in rat ventral prostate. Mol. Pharmacol. 5:428-431.

Forsberg, J.-G. and Kalland, J. (1981). Neonatal estrogen treatment and epithelial abnormalities in the cervicovaginal epithelium of adult mice. Cancer Res. 41:721-734.

Friedenwald, J. S. (1951). Growth pressure and metaplasia of conjunctival and corneal epithelium. Doc. Ophthalmol. 5:184-192.

Fujii, H., Cunha, G. R., and Norman, J. T. (1982). The induction of adenocarcinomatous differentiation in neoplastic bladder epithelium by an embryonic prostatic inductor. Invest. Urol. (submitted).

Higgins, S. J. and Parker, M. G. (1980). Androgenic regulation of generalized and specific responses in accessory sexual tissues of the male rat. In Biochemical Actions of Hormones (G. Litwak, ed.), pp. 287-309. Academic Press, New York.

Higgins, S. J., Purchell, J. M., and Mainwaring, W. I. P. (1976). Androgen-dependent synthesis of basic secretory proteins by the rat seminal vesicle. Biochem. J. 158:271-282.

Hodges, G. M., Hicks, R. M., and Spacey, G. D. (1977). Epithelial-stromal interactions in normal and chemical carcinogen-treated adult bladders. Cancer Res. 37:3720-3730.

Karkinen-Jaaskelainen, M. (1978). Transfilter lens induction in avian embryos. Differentiation 12:31-37.

Karring, T., Lang, N. P., and Loe, H. (1975). The role of gingival connective tissue in determining epithelial differentiation. J. Periodont. Res. 10:1-11.

Kollar, E. J. (1972). The development of the integument: spatial, temporal, and phylogenetic factors. Am. Zool. 12:125-135.

Kratochwil, K. and Schwartz, P. (1976). Tissue interaction in androgen response of embryonic mammary rudiment of mouse: Identification of target tissue of testosterone. Proc. Natl. Acad. Sci.
USA 73:4041-4044.

Lasnitzki, J. and Mizuno, T. (1980). Prostatic induction: Interaction
of epithelium and mesenchyme from normal wild-type and androgen-
insensitive mice with testicular feminization. J. Endoc. 85:423-
428.

Lawson, K. A. (1972). The role of mesenchyme in the morphogenesis
and functional differentiation of rat salivary epithelium. J. Embryol. Exp. Morphol. 27:497-513.

MacKenzie, J. C., Dabelsteen, E., and Roed-Peterson, B. (1979).
A method for studying epithelial-mesenchymal interactions in
human oral mucosa lesions. Scand. J. Dent. Res. 87:234-243.

Mainwaring, W. I. P. (1977). The Mechanism of Action of Androgens,
Springer-Verlag, New York.

McLachlan, J. A., Newbold, R. R., and Bullock, B. C. (1981).
Long-term effects on the female mouse genital tract associated
with prenatal exposure to diethylstilbestrol. Cancer Res. 40:
3988-3999.

Mizuno, T., Yasugi, S., and Sumiya, M. (1975). The determination,
cytodifferentiation and morphogenesis of the digestive tract epithelium. Dev. Growth, Differ. 17:315-316.

Ohno, S. (1977). The Y-linked H-Y antigen locus and the X-linked
Tfm locus as major regulatory genes of the mammalian sex determining mechanism. J. Steroid Biochem. 8:585-592.

Oliver, R. F. (1968). The regeneration of vibrissae. A model for
the study of dermal-epidermal interactions. In Epithelial-Mesenchymal Interactions (R. Fleischmajer and R. E. Billingham, eds.),
pp. 267-279. Williams & Wilkins, Baltimore.

O'Malley, B. W. and Schrader, W. T. (1976). The receptors of steroid hormones. Sci. Am. 234:32-43.

Parker, M. G., Scrace, G. T., and Mainwaring, W. I. P. (1978).
Testosterone regulates the synthesis of major proteins in rat
ventral prostate. Biochem. J. 170:115-121.

Sakakura, T., Nishizuka, Y., and Dawe, C. J. (1976). Mesenchyme-dependent morphogenesis and epithelium-specific cytodifferentiation in mouse mammary gland. Science 194:1439-1441.

Sakakura, T., Nishizuka, Y., and Dawe, C. J. (1979a). Capacity of mammary fat pads of adult C3H/HeMs mice are capable to interact morphogenetically with fetal mammary epithelium. J. Natl. Cancer Inst. 63:733-736.

Sakakura, T., Sakagami, Y., and Nishizuka, Y. (1979b). Acceleration of mammary cancer development by grafting of fetal mammary mesenchymes in C3H mice. Gan. 70:459-466.

Sakakura, T., Sakagami, Y., and Nishizuka, Y. (1981). Accelerated mammary cancer development by fetal salivary mesenchyme isografted to adult mouse mammary epithelium. J. Natl. Cancer Inst. 66:953-959.

Sengel, P. (1976). Morphogenesis of Skin. Cambridge University Press, New York.

Shannon, J. M., Cunha, G. R., and Vanderslice, K. D. (1981). Autoradiographic localization of androgen receptors on the developing urogenital tract and mammary gland. Anat. Rec. 199:232A.

Stumpf, W. E., Narbaitz, R., and Sar, M. (1980). Estrogen receptors in the fetal mouse. J. Steroid Biochem. 12:55-64.

Takasugi, N. (1976). Cytological basis for permanent vaginal changes in mice treated neonatally with steroid hormones. Int. Rev. Cytol. 44:193-224.

Wilson, J. D., Griffin, J. E., Leshin, M., and George, F. W. (1981). Role of gonadal hormones in development of the sexual phenotypes. Hum. Genet. 58:78-84.

Yasugi, S. and Mizuno, T. (1978). Differentiation of the digestive tract epithelium under the influence of the heterologous mesenchyme of the digestive tract in the bird embryo. Dev. Growth, Differ. 20:261-267.

# 4

# ROLE OF EXTRACELLULAR MATRIX IN CELL DIFFERENTIATION AND MORPHOGENESIS

## A. H. Reddi

INTRODUCTION

Cell differentiation and morphogenesis are the major problems confronting the contemporary biologist. A better understanding of the mechanisms of regulation of differential gene expression has immense implications for cellular and developmental biology and in the pathogenesis of cancer. While considerable attention has been focused on the role of nuclear and cytoplasmic macromolecules involved in differentiation, information concerning the extracellular matrix influences has lagged. This review will examine the recent advances in the biology and chemistry of collagens, proteoglycans, fibronectin, and other extracellular matrix constituents. Further, special emphasis will be placed on extracellular collagenous matrix-induced bone differentiations and morphogenesis. This system is a prototype for the study of matrix-cell interactions in vivo. Finally this brief review will conclude with a discussion of the role of the emerging model for cell-matrix continuum and its possible implications for specification of positional information during embryonic development and in tissue repair and regeneration.

ORIGIN AND EVOLUTION OF THE
EXTRACELLULAR MATRIX

A cursory examination of the phylogenetic relationships in the animal kingdom reveals that the origin and evolution of multicellular metazoa was accompanied by the appearance of extracellular matrix (Borradaile et al., 1958; Reddi, 1976). For example in the genus

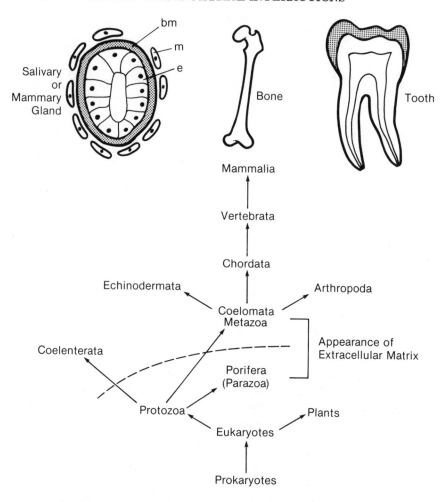

Figure 4.1 Phylogenetic aspects of the origin and evolution of extra-cellular matrix. Note the discrete appearance of extracellular matrix in multicellular metazoa. Bones and teeth abound in extracellular matrix.

Hydra the mesoglea represents a true extracellular matrix. Figure 4.1 summarizes the evolutionary aspects of the origin of an extra-cellular matrix. Prokaryotes such as Escherichia coli lack a discrete nucleus. The origin and evolution of unicellular eukaryotes such as yeast and amoeba were marked by the appearance of discrete nucleus. However, when one examines the evolution of multicellular metazoan eukaryotes starting from coelenterates to mammals the organization of tissue form reveals prominent extracellular matrices. In glandular

structures such as salivary glands or mammary glands the heterogeneous nature of the cells (epithelial and mesenchymal cells) and
the prominent basement membranes are evident. It is well known
that the epithelial-mesenchymal interactions are of paramount importance in tissue morphogenesis and the three-dimensional branching
pattern of diverse organs such as lungs, pancreas, and salivary
glands (Grobstein, 1967; Reddi, 1976). The classical studies of
Grobstein (1954; 1967; 1975) have laid the foundations for our current
ideas in this area. There are certain tissues in the mammals such
as teeth and bones which abound in extracellular matrices and are
invaluable aids to the students of extracellular matrix biology. The
dentin matrix in the tooth is essentially cell-free and is predominantly composed of collagenous matrix. Likewise, the bone matrix
which was synthesized and assembled by bone forming osteoblasts
is composed of a vast expanse of extracellular collagenous matrix
with osteocytes located in the matrix. In view of this, teeth and bone
matrices are useful choices for studies of the role of extracellular
matrices in cell differentiation, morphogenesis, and in repair and
regeneration of tissues.

## EXTRACELLULAR MATRIX CONSTITUENTS

The extracellular matrix of the various tissues in the multicellular eukaryotes is composed of collagens, proteoglycans, and a
variety of glycoproteins including fibronectins. It is conceivable that
at least some of these glycoproteins may be involved in the adhesion
of cells to extracellular matrix and may thereby constitute the cell-
matrix interface. We shall now describe the various components of
the extracellular matrix.

### Collagens

Collagens are a family of several structurally and functionally
related glycoproteins. In this respect they are similar to the multi-
gene families of proteins such as immunoglobulins and hemoglobins.
There are structurally distinct collagen types and their tissue distribution is summarized in Table 4.1. Generally, collagens consist of
a triple-helical conformation (Ramachandran and Reddi, 1976; Bornstein and Sage, 1980; Eyre, 1980) and have a characteristic amino
acid composition that has 33 percent glycine and a combined proline
and hydroxyproline of about 20-22 percent. The triple-helical collagen
molecule has a central helical domain and two nonhelical domains in
the N (amino) and C (carboxy) terminal regions of the molecule. The

TABLE 4.1

Tissue Distribution of Collagen Types

| Type | Chains | Molecular Stoichiometry | Tissue Distribution |
|------|--------|------------------------|---------------------|
| I | $\alpha 1$, $\alpha 2$ | $[\alpha 1\ (I)]_2\ \alpha 2\ (I)$ | Bone, skin, tendon, dentin in tooth |
| II | $\alpha 1$ | $[\alpha 1\ (II)]_3$ | Cartilage, notochord, vitreous body in eye |
| III | $\alpha 1$ | $[\alpha 1\ (III)]_3$ | Embryonic skin, blood vessels, uterus |
| IV | $\alpha 1$, $\alpha 2$ | $[\alpha 1\ (IV)]_3$ or $[\alpha 2\ (IV)]_3$ | Basement membranes |
| V | $\alpha 1$, $\alpha 2$, $\alpha 3$ (B, A, C respectively) | Stoichiometry not established | Capillaries, smooth muscle, placenta |

Source: Compiled by author.

monomeric collagen is self-assembled extracellularly into higher order structures such as microfibrils and fibrils giving rise to characteristic bands in negatively stained electron micrographs.

A monomeric collagen molecule has the dimensions of $1.5 \times 300$ nm and consists of three polypeptide chains of about 95,000 daltons. The various types of collagen molecules that have been characterized appear to be tissue specific, although it is not uncommon to notice considerable codistribution in several anatomical locations. Type I collagen is found in bone, skin, and tendon and has two (alpha) $\alpha 1$ (I) chains and one $\alpha 2$ (I) chain. Cartilage consists of mainly type II collagen and contains three $\alpha 1$ (II) chains. Type III collagen consists of three $\alpha 1$ (III) chains and is distributed in blood vessels, fetal skin, and numerous parenchymatous tissues such as liver, spleen, and bone marrow (Reddi et al., 1977). Type IV collagen is found in basement membranes and consists of two distinct chains. Type V collagen is present in smooth muscle, placenta, and in bone (Bornstein and Sage, 1980; Kleinman et al., 1981).

The biosynthesis of collagen occurs intracellularly and a biosynthetic precursor, procollagen has been isolated and characterized (Gross, 1974; Gay and Miller, 1978; Fessler and Fessler, 1978). This molecule has an amino terminal appendage of about 14,000 daltons and a carboxy terminal extension of 36,000 daltons. These peptides are cleaved by proteolytic processing extracellularly prior to fibril formation by self-assembly. The constituent chains of the triple-helix are extensively modified post translationally (Prockop et al., 1979). These modifications include: hydroxylation of proline and lysine, galactosyl and glucosyl transfer to certain hydroxylysines, oxidative deamination of lysine and hydroxylysine to yield cross-link precursors. The fully formed fibrils of collagen are generally resistant to most proteases. Degradation of collagen is initiated by a specific enzyme collagenase (Harper, 1980). Mammalian collagenase cleaves the native triple-helical monomer to two asymmetric pieces of three-quarters and one-quarter length. This results in unwinding of the triple helix with the resultant rapid degradation of the polypeptide by other proteases in the vicinity.

Collagen synthesis is detected very early during embryogenesis (Reddi, 1976). In fact collagen synthesis precedes the appearance of histologically distinct connective tissue cells. Collagenous proteins have been implicated in epithelial-mesenchymal interactions during morphogenesis of a variety of tissues (Grobstein, 1975). Although mesenchyme is generally considered to be the site of synthesis of collagen, there is a growing realization of the role of epithelial cells in collagen production (Reddi, 1976; Bornstein and Sage, 1980; Hay, 1981). The intricate mechanisms involved in the biosynthesis of collagen and its posttranslational modification indicates that any deficiencies in this process might result in disorders of collagen metabolism and this indeed is the case (Prockop et al., 1979; Minor, 1980).

Proteoglycans

Proteoglycans may be defined as macromolecules with a protein core to which glycosaminoglycans (acid mucopolysaccharides) are covalently attached. These macromolecules are ubiquitous in cartilage matrices and have been studied extensively (Hascall, 1981; Hascall and Heinegard, 1979; Roden, 1980; Muir and Hardingham, 1975; Rosenberg, 1975). The current model for cartilage proteoglycan aggregates is presented in Figure 4.2. The proteoglycans are extremely heterogeneous in their structure and consist of monomers (subunits) that interact noncovalently with hyaluronic acid (HA) to form large molecular weight aggregates. These interactions between monomer and hyaluronic acid are stabilized by "link" glycoproteins.

Each monomer in turn is composed of a core protein with a molecular weight of 250,000 daltons and extends about 300 nm in length. Glycosaminoglycan (GAG) chains are covalently attached to the protein core. In the rat cartilage proteoglycan GAG chains are predominantly chondroitin sulfate (CS) (Reddi et al., 1978) whereas in the bovine keratan sulfate is found in addition to the CS. The glycosaminoglycans are linear polymers of repeating disaccharides of hexuronic acid and hexosamine. In bovine cartilages there are 60–80 chondroitin sulfate chains with an average molecular weight of 20,000 daltons. Each chain is attached to a serine hydroxyl of the core protein by an

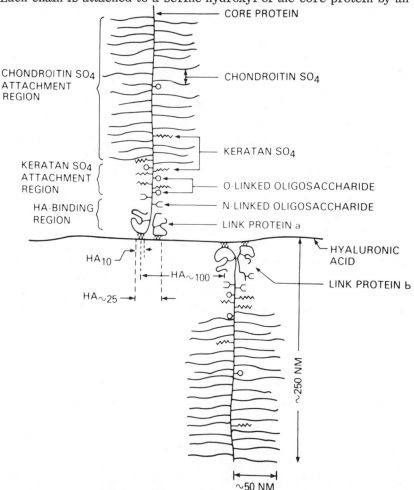

Figure 4.2 A model for cartilage proteoglycan structure. The aggregate is formed by noncovalent interaction of the monomer proteoglycan with hyaluronic acid. This interaction is stabilized by link glycoprotein. Reproduced with permission from Hascall (1981).

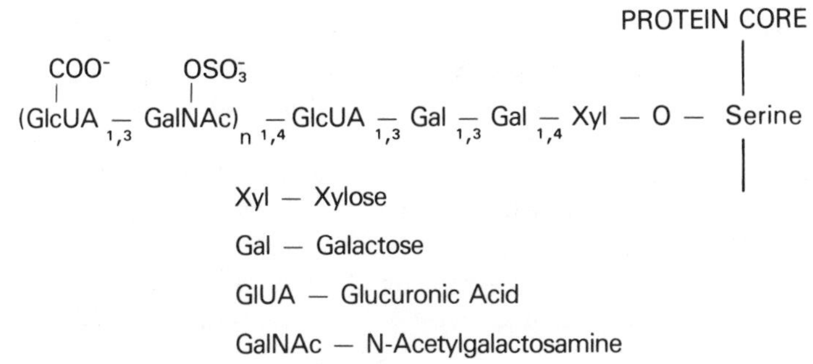

PROTEIN CORE

$$\text{(GlcUA} \underset{1,3}{-} \text{GalNAc)}_n \underset{1,4}{-} \text{GlcUA} \underset{1,3}{-} \text{Gal} \underset{1,3}{-} \text{Gal} \underset{1,4}{-} \text{Xyl} - \text{O} - \text{Serine}$$

COO⁻ over GlcUA, OSO₃⁻ over GalNAc

Xyl — Xylose

Gal — Galactose

GlUA — Glucuronic Acid

GalNAc — N-Acetylgalactosamine

Figure 4.3  The glycosaminoglycan (acid mucopolysaccharide) attachment region of the protein core in chondroitin sulfate proteoglycan.

O-glycosidic bond to the xylose residue of the reducing terminus of the chondroitin sulfate chain (Fig. 4.3). The linkage region is illustrated in Figure 4.3 and consists of two galactose residues and one glucuronic acid to which the repeating disaccharides galactosamine and glucuronic acid are attached. Keratan sulfate chains are attached to the core protein by O-glycosidic bonds between serine or threonine hydroxyls of protein and N-acetylglucosamine. The repeating disaccharide in keratan sulfate consists of N-acetylglucosamine and galactose. Certain N-linked oligosaccharides were recently described in the core protein (Hascall, 1981). The three major domains of the core protein are: HA binding region, the keratan sulfate attachment region, and chondroitin sulfate region. The HA binding region of the core protein interacts with a decasaccharide (five repeating disaccharides) and is rather specific. This emerging model for proteoglycan structure in cartilage has set the stage for studies on a variety of extracellular matrices of tissues such as aorta, skin, liver, and bone (Hascall, 1981; Roden, 1980).

The increasing knowledge of proteoglycan chemistry has laid the foundation for probing the biological role in development and morphogenesis. One approach has been to investigate (Kinoshita and Saiga, 1979; Kinoshita and Yoshii, 1979) the inhibitory influence of specific metabolic inhibitors such as aryl-$\beta$-D-xyloside (inhibitor of polysaccharide attachment to core protein), sodium selenate (inhibior of sulfation of the glycosaminoglycan), and 2-deoxy-D-glucose (inhibitor of chain elongation of glycosaminoglycan). These experiments revealed that sea urchin embryonic development was arrested as blastula stage hinting at the crucial requirement of optimal proteoglycan biosynthesis for further growth and morphogenesis. These developmental influences were stage specific and were alleviated by exogenous proteoglycans isolated from postgastrula stage sea urchins.

Judicious study of certain developmental disorders may shed light on the role of proteoglycans in development. This certainly was the case in the study of nanomelia, an autosomal recessive mutation (Pennypacker and Goetinck, 1976) in chicks and it was found that the biosynthesis of cartilage proteoglycans was severely impaired. However, the cartilage-specific type II collagen biosynthesis was not affected. It would appear that a defect in proteoglycan biosynthesis and assembly inhibits the longitudinal growth of cartilage (Pennypacker and Goetinck, 1976).

## Fibronectins and Other Glycoproteins

Fibronectins are high molecular weight cell surface glycoproteins involved in adhesion of cells to collagen and other components of extracellular matrix (Yamada and Olden, 1978; Vaheri et al., 1978; Pearlstein et al., 1980; Kleinman et al., 1981). The plasma fibronectins (cold insoluble globulins) are immunologically cross-reactive with the cell surface fibronectin and there are clearly structural similarities. This glycoprotein is a disulfide bonded dimer of a molecular weight of 450,000 daltons. The carbohydrate content ranges between 5 and 6 percent by weight. As seen from Table 4.2, the molecule has been implicated in cell-cell aggregation and cell-extracellular matrix adhesion. In addition the fibronectins have molecular domains with affinity for collagens, heparin, and fibrin (Fig. 4.4).

TABLE 4.2

Biological Functions of Fibronectin

---

1. Cell-cell aggregation

2. Cell-substratum adhesion

3. Binding to collagen, fibrin, and heparin

4. Nonspecific opsonization in reticuloendothelial system

5. Phenotype transformation of cells
   a) Transformed cells
   b) Chondrocytes

6. Inhibition of myoblast fusion.

---

Source: Compiled by author.

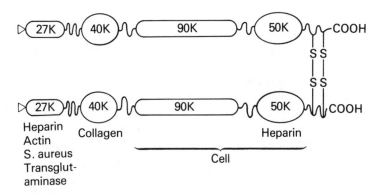

Figure 4.4 An evolving model for structural and functional domains in fibronectin. A disulfide-linked dimer of 220,000 dalton subunits is shown. The molecular weights of protease-resistant domains that have been purified are indicated above. The functional domains are indicated below. Reproduced by courtesy of Yamada (1982).

It is conceivable that these interactions may be crucial for early phases of wound healing, fracture repair, and regenerative growth phenomena in a variety of tissues including limbs. The term fibronectin (from fibra, fiber and nectere, to bind) is now widely accepted in the literature (Vaheri et al., 1978).

Fibronectins are distributed ubiquitously in a variety of cell types that include: endothelial cells, mesenchymal cells, myoblasts, chondroblasts, Schwann cells, and some cultured epithelia cells. Cellular fibronectin has been localized by indirect immunofluorescence and appears as fibrillar arrays on the cell surface and in numerous basement membranes. This glycoprotein is greatly reduced or lost on transformation of cells by viruses (Hynes, 1976). Many of the morphological changes due to virus-induced transformation can be partially reversed by the addition of purified cellular fibronectin (Yamada and Olden, 1978). Conversely, treatment of normal cells with antifibronectin antibodies results in alteration of cell shape reminiscent of that observed in transformed cells. Fibronectin functions as an $\alpha 2$-opsonic protein during phagocytosis and may be involved in wound debridement prior to the healing response (Saba et al., 1978). The plasma levels of fibronectin have been correlated with the phagocytic function of the reticuloendothelial system. In patients with trauma and burns there is a decrease in plasma fibronectin and a parallel decline in reticuloendothelial system function. Fibronectin appears ago to be a chemotactic molecule for fibroblasts (Gauss-Müller et al., 1980). A further role for fibronectin in wound healing and in embryonic development is suggested by the transglutaminase (Folk, 1980) catalyzed cross-linking of collagen and fibronectin (Mosher et al., 1979). The rapid advances in this realm of extra-cellular matrix biology may yield exciting results and unlock the secrets of matrix-cell interactions.

Laminin is a newly described noncollagenous glycoprotein of basement membrane matrix producing murine tumor and has been localized in several normal tissues in the basal lamina (Timpl et al., 1979). It has a molecular weight of about 1,000,000 daltons and it consists of at least two polypeptide chains of 200,000 daltons and 400,000 daltons. The exact molecular architecture is under intense scrutiny. Laminin has been implicated in the attachment of epithelial cells to basement membrane collagen (Terranova et al., 1980). Chondronectin is a glycoprotein implicated in chondrocyte attachment to substratum (Kleinman et al., 1981).

## EXTRACELLULAR MATRIX-INDUCED BONE DIFFERENTIATION

An interesting biological effect is elicited by subcutaneous implantation of demineralized, diaphyseal, extracellular bone matrix in the growing rat. The implanted extracellular matrix is predominantly collagenous in nature with some tightly associated glycoproteins (Reddi, 1976) and results in the sequential development of endochondral bone (Reddi and Huggins, 1972; 1975; Reddi and Anderson, 1976; Reddi, 1981). This biological cascade is described briefly in view of its implications for the role of extracellular matrix in development and morphogenesis (Fig. 4.5). On implantation of the extracellular matrix from bone there was an instantaneous formation of blood clot. On day 1, the implant was a button-like, plano-convex plaque with implanted matrix, fibrin, and neutrophils. On day 3 the mesenchyme proliferated as indicated by $^3$H-thymidine incorporation and radioautography and by ornithine decarboxylase activity (Rath and Reddi, 1979). The earliest chondroblasts were seen on day 5 and the implant was a conglomerate of chondrocytes and implanted matrix on days 7-8 as observed morphologically and by the incorporation of $^{35}$SO$_4$ into proteoglycans (Reddi et al., 1978) and type II collagen localization (Reddi et al., 1977). On day 9 calcification of the hypertrophic cartilage matrix occurred and vascular invasion was observed. Basophilic osteogenic cells and osteoblasts were seen on days 10-11; and new bone formation was evident as indicated by increases in alkaline phosphatase activity and $^{45}$Ca incorporation (Reddi, 1981). On days 12-18 there was extensive bone remodeling and the newly formed ossicle is the site of hematopoietic bone marrow

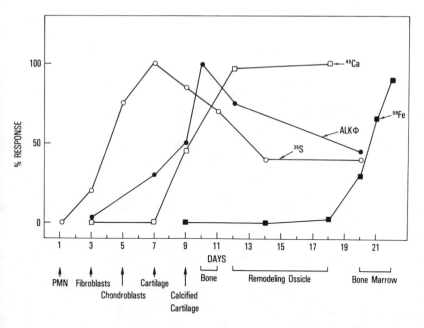

Figure 4.5 Sequential changes in $^{35}SO_4$ incorporation into proteoglycans, $^{45}Ca$ incorporation into mineral phase and $^{59}Fe$ incorporation into heme in response to subcutaneous implantation of extracellular bone matrix. Reproduced from Reddi (1981).

differentiation as monitored by $^{59}Fe$ incorporation into heme (Reddi and Huggins, 1975; Reddi, 1981). The response to the extracellular matrix is stringently specific and is elicited by only bone and tooth matrix and not by tendon and skin (Reddi, 1976).

The matrix-induced bone differentiation and morphogenesis is a prototype for the study of the role of extracellular matrix in development. This is probably the first known example of the potential of extracellular matrix to initiate cell differentiation in vivo. In view of this we have explored this experimental model to gain further insights into the role of matrix in morphogenesis.

The surface charge is critical. Alteration of the surface of the matrix by N-acetylation, carboxymethylation, and modification of the guanidino groups of arginine abolishes the inductive response. Also the pretreatment with heparin prevents cell differentiation (Reddi, 1976). The property of the matrix is sensitive to proteolytic treatment. Furthermore, the physical dimensions of the matrix have important bearing on the quantitative yield of bone. Matrix particles of less than 74 $\mu$m in size is feeble in its bone induction potential. This finding has important implications from the standpoint of

minimum dimensions of the substratum for the anchorage-dependent
cell proliferation (Stoker et al., 1968; Folkman and Moscona, 1978;
Gospodarowicz et al., 1980; Reddi, 1976).

Although the detailed molecular mechanisms are not known,
we have examined the role of fibronectin in early matrix-cell interactions in vivo (Weiss and Reddi, 1980; 1981). The implanted matrix
bound plasma fibronectin and may constitute an important initial
event for cell attachment to the matrix. The peak in fibronectin biosynthesis on day 3 coincides with the occurrence of type III collagen
in day 3 plaques (Steinmann and Reddi, 1980). This is interesting in
view of the finding that fibronectin binds avidly to native type III collagen in comparison to types I and II (Ruoslahti and Engvall, 1978;
Hormann and Jilek, 1978). Recent experiments revealed that 4.0 M
guanidine extracts of bone matrix are intensely chemotactic for rat
embryo fibroblasts (Reddi and Seppa, unpublished observations).
Recently the dissociatively extracted components from the extracellular matrix were fractioned by gel filtration on Sepharose 4B-CL and
fractions (< 50,000 in M.W.) when implanted after reconstitution
with collagenous bone matrix residue yielded new bone (Sampath and
Reddi, 1981). The same fractions were found to be mitogenic in vitro
to human and rat fibroblasts (unpublished observations). It is therefore likely that the extracellular matrix has chemotactic, mitogenic,
and differentiative functions and may function as a sequential cascade
analogous to the enzyme cascade in the blood coagulation system or
the immune complement system. Undoubtedly future experiments in
this area are likely to be informative, and may provide a molecular
basis for the mechanism of action of extracellular matrix components.


EMERGING MODEL FOR CELL-EXTRACELLULAR
MATRIX CONTINUUM

The growing knowledge of the extracellular matrix macromolecules such as collagens, proteoglycans, and fibronectins permits one
to construct models of the cell surface-extracellular matrix interface. One current working model is present in Figure 4.6. The
molecular details are not necessarily to scale. It emphasizes that
the cell surface-extracellular matrix is a continuum. It is difficult,
perhaps impossible to demarcate where a cell surface ends and the
extracellular matrix begins. The multiple interactions of the cell
surface fibronectin with extracellular matrix macromolecules such
as collagens, proteoglycans, heparin, and other glycosaminoglycans
(Yamada, 1982) and fibronectins of neighboring cells and matrix may
have important influence on restricting the mobility of cell surface/
membrane proteins, including receptors for hormones, growth

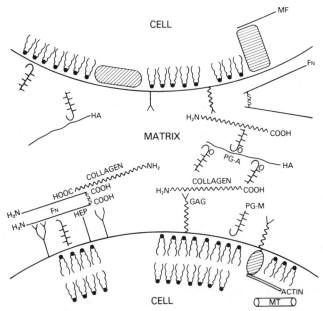

Figure 4.6 The emerging model for cell-extracellular matrix continuum. The cell surface/membrane glycosaminoglycans (GAG),
proteoglycan monomers (PG-M), heparin and heparan sulfates (HEP)
interact with fibronectins (Fn) and collagen. In addition to specific
domains for such interactions, electrostatic forces may also play a
role. The intracellular microfilaments (MF), microtubules (MT),
and actin cables may modulate and transduce external signals at the
cell surface-matrix interface. This might result in dynamic changes
in cell shape and locomotion as observed during embryonic development and morphogenesis.

factors, and a variety of developmental signals (as yet chemically
undefined). Also intracellular microtubules, microfilaments, and
actins may influence cell shape changes. It is well known that cell
membranes exhibit regional inhomogeneities and the polarized functions of cells such as secretion and virus budding are dependent on
substratum of cells (Rodriguez-Boulan, personal communication).
Furthermore, such local restriction of cell surface/membrane function by extracellular matrix components has far-reaching implications
for cell migration, cell sorting, and the interpretation of positional
information during embryonic development and morphogenesis. The
recent advances in the methodology of monoclonal antibodies and
recombinant DNA technology will help elucidate the molecular basis
for the role of extracellular matrix in development and morphogenesis.

REFERENCES

Bornstein, P. and Sage, H. (1980). Structurally distinct collagen types. Ann. Rev. Biochem. 49:957-1003.

Borradaile, L. A., Potts, F. A., Eastham, L. E. S., and Saunders, J. T. (1958). The Invertebrata (3rd ed., revised by G. A. Kerkut). Cambridge University Press, Cambridge.

Eyre, D. R. (1980). Collagen: molecular diversity in body's protein. Science 207:1315-1322.

Fessler, J. H. and Fessler, L. I. (1978). Biosynthesis of procollagen. Ann. Rev. Biochem. 47:129-142.

Folk, J. E. (1980). Transglutaminases. Ann. Rev. Biochem. 49: 517-531.

Folkman, J. and Moscona, A. (1978). Role of cell shape in growth control. Nature 273:345-349.

Gauss-Müller, V., Kleinman, H. K., Martin, G. R., and Schiffman, E. (1980). Role of attachment proteins and attractants in fibroblast chemotaxis. J. Lab. Clin. Med. 96:1071-1080.

Gay, S. and Miller, E. J. (1978). Collagen in the Physiology and Pathology of Connective Tissue. Gustav-Fischer Verlag, Stuttgart.

Gospodarowicz, D., Vlodavsky, I., and Savion, N. (1980). The extracellular matrix and the control of proliferation of vascular endothelial and vascular smooth muscle cells. J. Supramol. Struct. 13:339-372.

Grobstein, C. (1954). Tissue interaction in the morphogenesis of mouse embryonic rudiments in vitro. In Aspects of Synthesis and Order in Growth (D. Rudnick, ed.), pp. 233-256, Princeton University Press, Princeton.

Grobstein, C. (1967). Mechanisms of organogenetic tissue interaction. Natl. Cancer Inst. Monogr. 26:279-299.

Grobstein, C. (1975). Developmental role of intercellular matrix: retrospective and prospective. In Extracellular Matrix Influences on Expression (H. C. Slavkin and R. C. Greulich, eds.), pp. 9-16, Academic Press, New York.

Gross, J. (1974). Collagen Biology: Structure, Degradation and Disease. In Harvey Lectures, Vol. 68, pp. 351-432, Academic Press, New York.

Harper, E. (1980). Collagenases. Ann. Rev. Biochem. 49:1063-1078.

Hascall, V. C. (1981). Proteoglycans: Structure and Function. In Biology of Carbohydrates (V. Ginsburg, ed.), Vol. 1, pp. 1-49, John Wiley and Sons, New York.

Hascall, V. C. and Heinegard, D. (1979). Proteoglycans. In Glyco-conjugate Research (J. D. Gregory and R. W. Jeanloz, eds.), pp. 341-374, Academic Press, New York.

Hay, E. D. (1981). Extracellular matrix. J. Cell Biol. 91:205S-223S.

Hormann, H. and Jilek, F. (1978). Cold insoluble globulin. Trans-amidation sensitive site and affinity behavior. Ann. NY Acad. Sci. 312:399-400.

Hynes, R. O. (1976). Cell surface proteins and malignant transformation. Biochim. Biophys. Acta 458:73-107.

Kinoshita, S. and Saiga, H. (1979). The role of proteoglycan in the development of sea urchins. I. Abnormal development of sea urchin embryos caused by the disturbance of proteoglycan synthesis. Exp. Cell Res. 123:229-236.

Kinoshita, S. and Yoshii, K. (1979). The role of proteoglycan synthesis in the development of sea urchins. II. Effect of administration of exogenous proteoglycan. Exp. Cell Res. 124:361-369.

Kleinman, H. K., Klebe, R. J., and Martin, G. R. (1981). Role of collagenous matrices in the adhesion and growth of cells. J. Cell Biol. 88:473-485.

Minor, R. R. (1980). Collagen metabolism. Ann. J. Pathol. 98:226-278.

Mosher, D. F., Schad, P. E., and Kleinman, H. K. (1979). Cross-linking of fibronectin to collagen by blood coagulation factor XIIIa. J. Clin. Invest. 64:781-787.

Muir, H. and Hardingham, T. E. (1975). Structure of proteoglycans. In Biochemistry of Carbohydrates (W. J. Whelan, ed.), pp. 153-222, Butterworth, London.

Pearlstein, E., Gold, L. I., and Garcia-Pardo, A. (1980). Fibronectin: a review of its structure and biological activity. Mol. Cell. Biochem. 29:103-128.

Pennypacker, J. P. and Goetinck, P. F. (1976). Biochemical and ultrastructural studies of collagen and proteochondroitin sulfate in normal and nanomelic cartilage. Dev. Biol. 50:35-47.

Prockop, D. J., Kivirikko, K. I., Tuderman, L., and Guzman, N. A. (1979). The biosynthesis of collagen and its disorders. New Engl. J. Med. 301:13-23 and 77-85.

Ramachandran, G. N. and Reddi, A. H. (1976). Biochemistry of Collagen, Plenum Press, New York.

Rath, N. C. and Reddi, A. H. (1979). Collagenous bone matrix is a local mitogen. Nature 278:855-857.

Reddi, A. H. (1976). Collagen and cell differentiation. In Biochemistry of Collagen (G. N. Ramachandran and A. H. Reddi, eds.), pp. 449-478, Plenum Press, New York.

Reddi, A. H. (1981). Cell biology and biochemistry of endochondral bone development. Collagen Res. 1:209-226.

Reddi, A. H. and Anderson, W. A. (1976). Collagenous bone matrix-induced endochondral ossification and hemopoiesis. J. Cell Biol. 69:557-572.

Reddi, A. H., Gay, R., Gay, S., and Miller, E. J. (1977). Transitions in collagen types during matrix-induced cartilage, bone and bone marrow formation. Proc. Nat. Acad. Sci. USA 74:5589-5592.

Reddi, A. H., Hascall, V. C., and Hascall, G. K. (1978). Changes in proteoglycan types during matrix-induced cartilage and bone development. J. Biol. Chem. 253:2429-2436.

Reddi, A. H. and Huggins, C. B. (1972). Biochemical sequences in the transformation of normal fibroblasts in adolescent rats. Proc. Natl. Acad. Sci. USA 69:1601-1605.

Reddi, A. H. and Huggins, C. B. (1975). Formation of bone marrow in fibroblast transformation ossicles. Proc. Nat. Acad. Sci USA 72:2212-2216.

Roden, L. (1980). Structure and metabolism of connective tissue proteoglycans. In Biochemistry of Glycoproteins and Proteoglycans (W. A. Lennarz, ed.), pp. 267-371, Plenum Press, New York.

Rosenberg, L. C. (1975). Structure of proteoglycans. In Dynamics of Connective Tissue Macromolecules (P. M. C. Burleigh and A. R. Poole, eds.), pp. 105-128, Elsevier, New York.

Ruoslahti, E. and Engvall, E. (1978). Immunochemical and collagen-binding properties of fibronectin. Ann. NY Acad. Sci. 312:178-191.

Saba, T. M., Blumenstock, J. A., Weber, P. A., and Kaplan, J. E. (1978). Physiological role of cold insoluble globulin in systemic host defence: Implications of its characterization as the $\alpha$2-surface binding glycoprotein. Ann. NY Acad. Sci. 312:43-55.

Sampath, T. K. and Reddi, A. H. (1981). Dissociative extraction and reconstitution of extracellular matrix components involved in local bone differentiation. Proc. Nat. Acad. Sci. USA 78:7599-7603.

Steinmann, B. U. and Reddi, A. H. (1980). Changes in synthesis of types I and III collagen during matrix-induced endochondral bone differentiation in rat. Biochem. J. 186:919-924.

Stoker, M., O'Neill, C., Berryman, S., and Waxman, V. (1968). Anchorage and growth regulation in normal and virus transformed cells. Int. J. Cancer 3:683-689.

Terranova, V. P., Rohrbach, D. H., and Martin, G. R. (1980). Role of laminin in the attachment of PAM 212 (epithelial) cells to basement membrane collagen. Cell 22:719-726.

Timpl, R., Rohde, H., Robey, P. G., Rennard, S. I., Foidart, J. M., and Martin, G. R. (1979). Laminin-a glycoprotein from basement membranes. J. Biol. Chem. 254:9932-9937.

Vaheri, A., Ruoslahti, E., and Mosher, D. F. (1978). Fibroblast surface proteins. Ann. NY Acad. Sci. 312:1-456.

Weiss, R. E. and Reddi, A. H. (1980). Synthesis and localization of fibronectin during collagenous matrix-mesenchymal cell interaction and differentiation of cartilage and bone in vivo. Proc. Nat. Acad. Sci. USA 77:2074-2078.

Weiss, R. E. and Reddi, A. H. (1981). Role of fibronectin in collagenous matrix-induced mesenchymal cell proliferation and differentiation in vivo. Exp. Cell Res. 133:247-254.

Yamada, K. M. (1982). Biochemistry of fibronectin. In The Glycoconjugates (M. I. Horowitz, ed.), Academic Press, New York.

Yamada, K. M. and Olden, K. (1978). Fibronectins-adhesive glycoproteins of cell surface and blood. Nature 275:179-184.

# 5

# THE ROLE OF DEFINED EXTRACELLULAR MATRICES ON THE GROWTH AND DIFFERENTIATION OF MAMMALIAN STRATIFIED SQUAMOUS EPITHELIUM

*John H. Lillie, Donald K. MacCallum, and Arne Jepsen*

The literature of developmental biology is replete with examples of epithelial-mesenchymal interactions. The intricacies of some of these interactions have been examined and reviewed during the previous chapters. The instructive element(s) involved in each of the interactions vary widely, but can be broadly grouped into three categories: 1) cell and cell surfaces; 2) diffusible cell products; and 3) nondiffusible cell products. The extracellular matrix, a mix of collagenous proteins and other associated proteins and glycoproteins, is generally considered to be an example of the latter. Some investigators, such as Reid and Rojkind (1979), have removed elements of the first two categories and studied the differentiative capacities of the residual extracellular matrices. Others have isolated specific elements of the matrix and studied the effect of each on the differentiation of cells in culture. Perhaps because collagens are generally the supporting framework to which other elements of the matrix are related, the interactions of various cells with different collagen types has been a logical first step in understanding the extracellular matrix. Studies by Meier and Hay (1973) and Emerman and Pitelka (1977) are examples of research which suggest that type I collagen directs the differentiation of various epithelia. Konigsberg and Hauschka (1965), Kosher and Church (1975), and a number of papers by Reddi (e.g., Reddi and Anderson, 1976) have shown that type I collagen in various forms influences the differentiation of a number of connective tissue cell types. Though dried films of type IV collagen have been shown to enhance cell differentiation of mammary epithelium (Wicha et al., 1979), studies on the influence of collagens other than type I have been limited.

A systematic, component by component study of the interactions of the extracellular matrix has not been considered possible for some tissues. For example, certain fastidious epithelia appear to require the presence of stromal fibroblasts or their secretions as a pre- requisite to culture. Such has been the case with stratified squamous epithelium (SSE) (Karasek and Charlton, 1971 and Rheinwald and Green, 1975). This stromal "requirement," itself an example of dependent interaction, has made investigation of the interactive role of specific matrix components difficult. But culturing SSE derived from the minimally keratinized, ventral surface of the rat tongue at low temperatures (32-34°C) (Jepsen, 1974), we have established a line of cells which is free of fibroblasts and is both phenotypically and karyotypically stable. These cells have been repeatedly passaged and can be subcultivated onto a variety of substrata (Jepsen et al., 1980). When cultured on type I collagen rafts at a gas-liquid inter- face, the cells organize into a well ordered SSE which can be main- tained for a prolonged period of time (30 d) in culture (Lillie et al., 1980).

Two of the three interstitial collagens, types I and III, have been isolated from the dermis (Miller et al., 1971). Immunofluores- cent studies of the distribution of the two collagens have shown that type III collagen antiserum binds most intensely to a narrow zone of the juxta-epithelial dermis, the superficial or papillary dermis (Gay et al., 1978). Located in this region in the normal cat, for example, is a fine meshwork of collagen fibrils which have a uniform diameter of approximately 76 nm. The cat reticular dermis, in contrast, is constructed of large (2-10 μm) randomly oriented fibers which are in turn composed of fibrils having diameters which are comparable to those of the superficial dermis (Patterson and Minor, 1977). The large fibers of this region appear, immunologically, to be composed of type I collagen (Gay et al., 1978). Interposed between the epithelium and this fibrous matrix is the basal lamina. In the studies I will pre- sent here, we have examined the interactions between SSE cells and each of these three elements of the extracellular matrix.

Type I collagen was extracted from guinea pig dermis by the method of Wahl et al. (1974). Differential salt precipitation of the extract (Kang et al., 1966) provided a preparation of relatively pure type I collagen. Type III collagen, in both the fully processed and pro-collagen forms, was prepared from fetal bovine skin using a neutral salt extraction technique (Lapiere et al., 1977). All tissue preparations and extractions were carried out at 4 to -120°C and in the presence of the protease inhibitors, phenylmethyl sulfonyl fluoride (PMSF) and N-ethylmaleamide (NEM). The purity of the preparations was determined using sodium dodecyl sulfate-polyacrylamide gel electrophoresis (SDS-PAGE) following the method of Laemmli (1979), but modified by the addition of urea to allow the differential separation of $\alpha_1$ (I), $\alpha_2$ (I), $\alpha_1$ (III), and p$\alpha_1$ (III) (Hayashi and Nagai, 1979). Samples were run in both the reduced (dithiothreital DDT) and nonreduced condition to demonstrate the presence of interchain disulfide bonding, a unique feature of type III collagen (Lenaers and Lapiere, 1975; Timpl et al., 1975). Identity of the pro-collagen (III) was made by showing that limited digestion with pepsin (Rhodes and Miller, 1978) converted it into $\alpha_1$ (III) collagen as determined by subsequent PAGE.

Collagen gels were formed by thermal precipitation (Gross and Kirk, 1958) from solutions containing 0.6 mg or 0.1 mg of collagen/ ml in 0.2 M NaCl and 0.05 M Tris HCl (pH 7.5). The kinetics of the gelation process were used to further characterize the collagens (Lapiere et al., 1977) and to ensure the reproducibility of fibrilogenesis between collagen batches.

A basement membrane for use as a culture substratum was deposited in vitro on the surface of various collagen gels and plastic by bovine endothelial cells (Lillie and MacCallum, 1977; MacCallum et al., 1982). Cells were removed from the surface of the membrane using a solution of 0.5 percent sodium deoxycholate (DOC) in warm (37°C), distilled water (Carlson et al., 1978).

The surface appearance of the substrata used in these studies is shown in Figure 5.1. From these comparably magnified scanning electron micrographs (x 1000), one can readily appreciate the differences. The collagen preparations differed physically in two interrelated ways, the apparent fibril size and fibril arrangement. As reported earlier by Lapiere et al. (1977), we found that gels of type III collagen (Fig. 5.1c) were composed of uniformly sized 53.7 ± 7 nm diameter) fibrils which were randomly oriented to form a dense, even meshwork. While the individual fibrils of the type I collagen gel were similar in diameter to those of the type III gel (67.7 ± 0.3 nm), the distribution of fibrils in the type I collagen gel was very uneven* (Fig. 5.1a). Fibrils were organized into a coarse network of interconnecting, wheat sheaf-like bundles. In the center of these bundles individual fibrils were partially fused with others to form large, 250–400 nm fibrils. Because such fusions were multiple and occurred at random throughout the region, the fibrils were effectively organized into yet larger units which were roughly 3–10 $\mu$m in diameter. This type of fibrillar fusion and formation of larger

---

*Measurements of fibril diameter and evidence of lateral fusions in the type I collagen are based on TEM of glutaraldehyde-OsO$_4$ fixed samples with tannic acid mordanting.

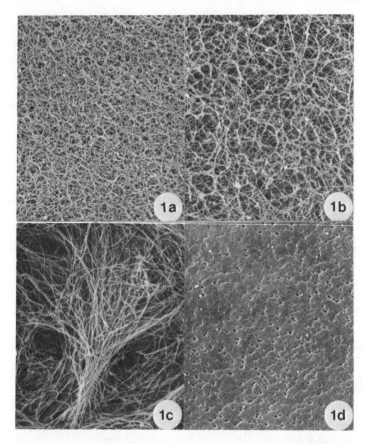

Figure 5.1a–d  Scanning electron micrographs of the surfaces the substrata used in these studies. Figures 5.1a–c are of collagen gels, types III, III, and I (3:7), respectively. The collagen concentration for each was 6 mg/ml. Figure 5.1d is of a basement membrane deposited on the surface of a type III collagen gel by bovine corneal endothelial cells in culture. (x 1000)

collagen units has not been observed in normal tissues and should not be confused with the processes which lead to the formation of collagen fibers in vivo.* Though we have not observed fibrillar fusion in

---

*Minor has observed this type of partial fibrillar fusion in the dermis of cats (Patterson and Minor, 1977) and dogs (Minor, 1980) which have hereditarily defective fibrillogenesis.

commercial preparations of type I collagen, the architecture described for gels of both collagen types was consistent. This consistency was probably due to the careful control of gelatin conditions and the use of collagen preparations which were relatively free of other collagen types and in which nonspecific proteolysis had been controlled. Both Helseth and Veis (1981) and Fessler et al. (1981) have shown that the protease sensitive, nonhelical telopeptides of both of these collagen types, as well as the procollagen extension of type III collagen direct or otherwise influence fibril formation in vitro. Furthermore, the addition of small amounts of type III collagen to the type I preparation greatly reduced the number of fibrillar fusions and resulted in gels with an architecture which most closely resembled the pure type III collagen gel. An example of a gel composed of both I and III collagen types (7:3 respectively) is shown in Figure 5.1b.

Direct comparisons between the gel substrata and the dermis are probably not warranted, but some similarities do exist. The juxta-epithelial dermis is, for example, predominantly composed of type III collagen. The fibrils of the region are comparable in size to those of the type III collagen gel and like them, are organized to form a fine meshwork rather than larger collagenous units. In contrast, the large fibers of the reticular dermis are composed for the most part of type I collagen and, like the wheat sheaf bundles of the gels, are rather widely spaced throughout the deep dermis to form the typical, relatively coarse network of the region.

The surface of the cleaned basement membrane (Fig. 5.1d) is pitted; a result of the lysis and dissolution of endothelial cell processes which were trapped in the basement membrane during its deposition. When deposited on type III collagen (as it is here) or on plastic, the membrane is flat and smooth. When deposited on type I collagen (Fig. 5.6) the surfaces of the membrane reflects the irregular contours of the underlying gel.

In preparing the cells for these experiments we had to consider the differential heterogeneity of SSE. Because only basal and juxtabasal cells of the SSE are capable of attachment and division, a population enriched for these cells was used for attachment studies and studies in which the number of cells used to initiate matched cultures had to be closely controlled. The populations were prepared from confluent, vigorously keratinizing cultures by preliminarily passing the cells at a 1:4 surface to surface ratio 72 hours prior to subcultivating them onto the experimental substrata. Nonattached cells and debris from the initial separation were removed by repeated rinses 24 hours following the initial passage. Cells used in the attachment assay were labeled for 20 hours with [$^3$H] thymidine ($^3$H-T; 4 $\mu$Ci/ml) beginning 48 hours following the preliminary passage.

Attachment of $^3$H-T labeled cells to the various substrata was studied over a 6 hour period. The substrata were formed in Costar Multiwell Plates which provided a standard 2 cm$^2$ surface for cell attachment. Cells were suspended in minimum essential medium with Earle's salt mixture (MEM) buffered with Good's buffers (pH 7.4) and plated in the presence or absence of 5 percent bovine calf serum. At selected times following plating, cells were dislodged from the surface of the substrata by controlled pipetting. The attached cells and substrata were then put into solution by incubating them at 32°C overnight in 0.5 N NaOH containing 1 percent SDS. The resulting solution was neutralized, added to an aqueous counting scintilant and the amount of radioactivity determined by liquid scintillation counting (LSC). Attachment of cells was expressed as a percentage of the original $^3$H dpm bound to a standard surface area. Because the substrata varied in terms of surface coarseness, attachment data were corrected for physical cell trapping. To determine this, glutaraldehyde fixed cell suspensions were used in place of viable cells in the attachment assay. (Residual free aldehydes attached to the cell surfaces following fixation were quenched prior to plating by rinsing the cells in buffered glycine.)

The attachment of SSE basal cells to the two collagens was essentially the same except for the physical trapping of cells (13.4 percent) within the surface interstices of the type I gel. Attachment progressed slowly and was considerably less efficient than to tissue culture plastic or the basement membrane. Cell attachment to the latter was very rapid; 56 percent of the cells were adherent within the first 30 minutes and 85 percent by the end of 6 hours. The studies on which Figure 5.2 is based were made in the absence of serum. Preliminary studies showed that in the presence of 5 percent serum, attachment to tissue culture plastic was initially considerably lower and was the same or only slightly reduced for the other substrata (Lillie et al., 1981). It should be noted that after 3 hours the cell attachment in the absence of serum did plateau or, as in the case of the collagens, drop slightly. We feel, however, that this occurs after the attachment event and probably reflects a serum based nutritional requirement of the cells. The serum independent attachment of these cells agrees with the observations of two other groups, but on different bases. Terranova et al. (1980) have reported that, unlike the attachment of mesenchymally derived cells (Klebe, 1974), serum fibronectin is not required for epithelial cell attachment to collagen. On the other hand, Schor and Court (1979) have shown that the condition of the collagen is the critical factor. They report that both epithelial and mesenchymal cells are able to attach to native collagen in the absence of serum.

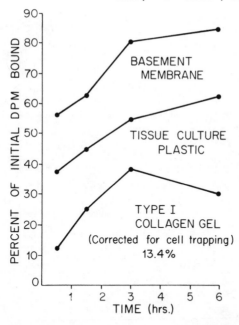

Figure 5.2 Attachment of [³H] thymidine labeled SSE cells to basement membrane, tissue culture plastic and gels of type I collagen
in the absence of serum. A factor (13.4 percent) correcting for the
physical trapping of the cells has been used to determine the collagen
curve. Each point represents the mean for 12 samples determined
during 3 separate experiments.

The behavior of cells during preconfluent growth was studied
using cultures which were initiated with matched inocula of the basal
cell rich preparation. During these studies the substrata were attached to the plastic culture surfaces. For study of long-term interactions (5-60 d) the epithelium was cultured using a raft technique.
The reason for this change, as we have previously reported (Lillie
et al., 1980), was that as post-confluent culture progressed, the
development of an extensive stratum corneum appears to compromise
the nutrition of the underlying basal cells. We have been able to sig-
nificantly prolong the culture period and promote SSE organization
by providing basal perfusion of the postconfluent epithelial sheet.
Therefore, the substratum, bearing the attached, newly confluent
epithelium was released from the surface on which it was formed,
floated onto a stainless steel grid, and was supported at the gas-
liquid interface in a conventional organ culture dish.

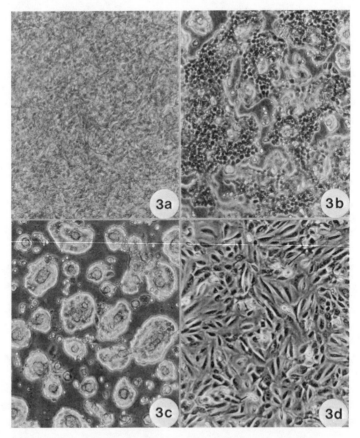

Figure 5.3a-d Illustrated in this set of phase micrographs is the colony morphology and outgrowth patterns of matched inocula of SSE cells grown 4 days on collagen gels (6 mg/ml) of type I (a), type III (c), and a mixture, 7:3 respectively, of the two collagen types (b), and on a basement membrane (d). (x 80)

The processes of cell migration, colony formation, and horizontal and vertical growth of the colonies, as first described by Karasek and Charlton (1971) for epithelial cells grown on type I collagen, varied considerably on the four substrata. The morphology of matched four day old cultures is shown in Figure 5.3. On the basement membrane (Fig. 5.3d), migration and colony formation was very rapid and the subsequent growth of cells was nearly exclusively horizontal. As a result, a minimally stratified confluent sheet of extensively flattened cells was formed. When grown on collagen, the basal-most cells were never as flattened as on the membrane and they migrated considerably more slowly. Cell behavior on the two collagens differed most during colonial outgrowth. Cells growing on type I collagen reached confluency first, but the epithelial sheet was only minimally keratinized (Fig. 5.3a). Colonies formed on type III collagen were numerous, but small in size and sharply circumscribed. Growth in these colonies was almost exclusively vertical and resulted in the formation of the keratinized cones shown in Figure 5.3c. A pattern of outgrowth which was intermediate between those described for the parent collagens was observed for colony

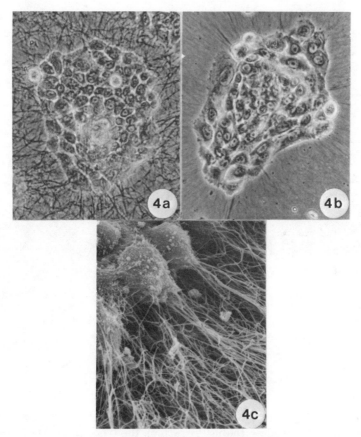

Figure 5.4a–c The ability or inability of SSE cells to orient the fibrils of collagen gels (1 mg/ml) of type I (a) or type III (b and c) is shown in this series. In the phase micrographs a and b (x 200), a radiating pattern of fibrils can only be seen surrounding the colony grown on the type III collagen. SEM of the periphery of the colony shown in b illustrates two cells attached to oriented bundles of collagen fibrils. (x 1170)

CARL A. RUDISILL LIBRARY
LENOIR RHYNE COLLEGE

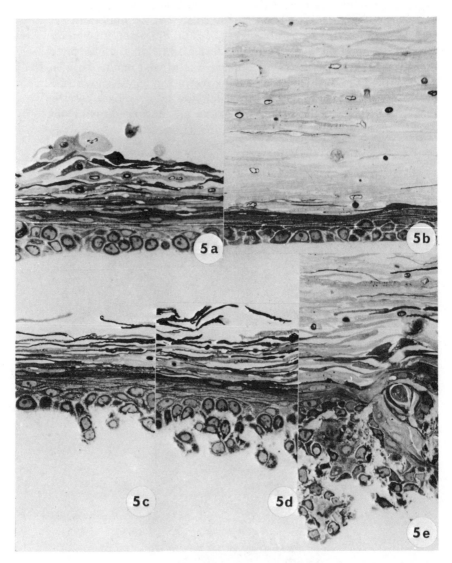

Figure 5.5a-e  In this series of light micrographs the maintenance
(12-50 d) of ordered epithelial growth on the type III collagen gel (a
and b) is contrasted to the sequential (12 d, c; 40 d, d; and 50 d, e)
loss of epithelial organization which is engendered by the disruption
of the epithelial-substratum interface. Both this figure and Figure
5.1 are magnified by x 1000. By directly comparing these two figures, the degree of support afforded to a single cell by each of the
substrata can be appreciated.

growth on the mixed collagen gels. They did, however, most resemble those formed on type III collagen (Fig. 5.3b).

Outgrowth on type III collagen was delayed by an intervening contractile event. When grown on gels having a low collagen concentration (1 mg/ml), it can be readily observed that the cells were able to influence the orientation of the individual type III fibrils (Fig. 5.4b), but were unable to affect the orientation of the coarser type I collagen bundles (Fig. 5.4a). Prior to migrating across the gel, the cells appeared to attach to the fibrils and draw them toward and under the colony (Fig. 5.4c). As a result, a dense fibrous mat is built up beneath the cells on the surface of the type III gel. This thin, cell-organized layer is even more complete than the original mesh work.

Others (Bell et al., 1979; Stopak and Harris, 1980) have shown that fibroblasts grown on, or in, gels of type I collagen can affect radial orientation of the fibril bundles. More recently, Harris et al. (1981) and Bellows et al. (1981) have shown that cell types vary in their ability to exert traction. For example, epithelial cells are unable, unless working in concert (Emerman and Pitelka, 1977), to influence a substratum composed of large bundles of type I collagen fibrils. But how likely is it that epithelial cells would ever interact with such a coarse matrix in vivo? During the processes of morphogenesis and wound healing, at times when the basal lamina has not yet formed or has been physically disrupted, it is much more likely that the cells would contact a fine fibrillar mesh, one which is typically composed of type III collagen. Tissue structure and our observations of cell growth on type III collagen suggest that the contractile strengths of cells may be appropriately matched to their substrata.

The importance of such support to the organization and maintenance of SSE in culture is shown in Figure 5.5. The alignment of basal cells on the surface of the type III collagen gel is well established by 12 d (Figure 5.5a) and maintained through 50 d in culture (Fig. 5.5b). As cells grow to confluence on type I collagen, some of the cells enter and line the surface irregularities of the gel (Fig. 5.5c; 12 d). Continued division and migration of these cells (Fig. 5.5d; 40 d) results in the disruption of the interface and the formation of basal cell nests. SSE maturation along this highly irregular surface and within the extensions leads to intraepithelial keratinization and cyst formation. The continuation of these two processes eventually affects the disorganization of the epithelium (Fig. 5.5e; 50 d).

While the disruptive effect of basal cell misalignment may be initially prevented by the formation of a dense, subepithelial mat, such support is probably not the means by which epithelial organization is maintained in vivo. When SSE cells are subcultured onto the surface of a highly irregularly contoured basement membrane, such as is deposited in the surface of the type I collagen gel, the cells attach rapidly, flatten, and migrate across the membrane. As they move, they trace out every surface irregularity and extension (Fig. 5.6). However, unlike the analogous situation on type I collagen, tissue organization is not destroyed by subsequent aberrant keratinization (Fig. 5.6; 50 d). It appears that once attachment and migration are completed on the basement membrane, the cells become quite stable. Evidence of the stability is best seen at the cell-basement membrane interface. In contrast to the rather pacific behavior of embryonic corneal epithelium on collagen (Meier and Hay,

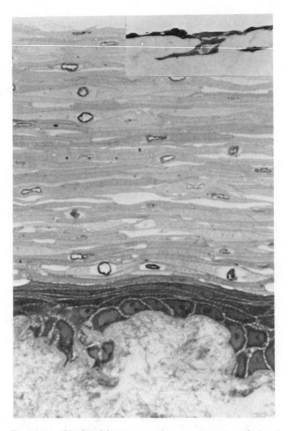

Figure 5.6 Despite the highly irregular contours of the basement membrane when it is deposited on type I collagen gels, epithelial order is maintained over 50 d in culture. (x 1504) The basal epithelial contour is established early in these cultures (Inset; 4 d; x 512). Note the sharply angled basal cell contours. This morphology is typical of SSE cells grown on this substratum.

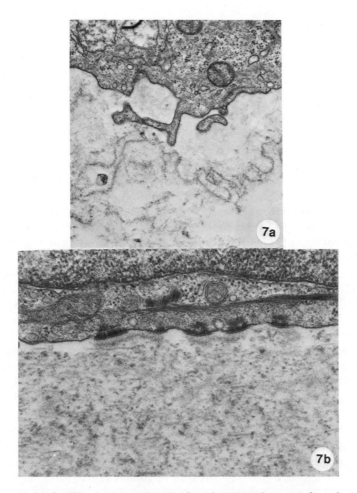

Figure 5.7a-b  Electron micrographs showing the interface between cells grown on type III collagen (a; x 23,200) and basement membrane (b; x 60,000). An electron dense ribbon of material is deposited on each substratum, but only against the basement membrane is the material deposited in a manner which is morphologically consistent with the formation of an epithelial basal lamina.

1973; 1974), the basal cells of SSE when grown on gels of either type I or III collagen continually extend and withdraw cell processes. A record of the changing interface is provided by the contours of an electron dense material which is deposited by the cells during culture (Fig. 5.7a). Against the basement membrane on the other hand (Fig. 5.7b), the basal plasma lemma becomes smooth and along its surface well-formed hemi-desmosomes appear. Opposite those attachment specializations and at a distance which approximates the width of the lamina lucida, the previously mentioned radio-dense material first appears. This pattern of deposition is consistent with that described by Briggaman et al. (1971) for the reappearance of the basal lamina following the recombination of the epidermis with the dermis. Whether, as reported by Sugrue and Hay (1981), the smoothing of the basal cell surface can be related to the release of soluble components of the matrix is not clear. However, since the membranes used in these studies were extensively washed in DOC, distilled water, and serum-free medium before use in culture, it is unlikely that such factors could have played any major role. (As previously noted, these two epithelial models may not be behaviorally comparable.)

Our findings, when considered along with the distribution of type III collagen in the dermis (Gay et al., 1978) and the relative increases of that collagen during development (Epstein, 1974) and wound healing (Miller, 1976), suggest that type III collagen may uniquely facilitate the establishment and initial maintenance of the epithelial sheet. While this function may appear to be later preempted by the synthesis and organization of the basal lamina, recent findings suggest a continuing role for type III collagen.

David and Bernfield (1981) have shown that type I collagen reduces the degradation of basal lamina proteoglycans. Other observations suggest that this function may not be unique to type I collagen and in fact may be even better served by type III collagen. Junqueira et al. (1981) have reported that the distribution of sulfated glycosaminoglycans among tissues having a predominant interstitial collagen type differs. They show that heparan sulfate characteristically distributes with type III collagen. Iterestingly, it is that particular glycosaminoglycan that Kanwar and Farquhar (1979) have isolated from renal basement membranes and Hassel et al. (1980) have shown to be distributed in the basement membranes of skin and cornea. Thus, as David and Bernfield originally proposed for type I collagen, type III collagen may effectively extend its initial influence over epithelial organization by stabilizing the turnover of laminar components.

ACKNOWLEDGEMENTS

The authors wish to thank Dr. Ronald R. Minor for his critical review of this manuscript. His observations regarding dermal architecture were most helpful. The fine technical assistance of Mr. Steve McKelvey in the preparation of the substrata and Ms. Angela Welford and Mr. Joe O'Brien in the preparation of material for electron microscopy and photographic printing are also gratefully acknowledged. Ms. Amy Pellegrino of the Department of Pathology in the College of Veterinary Medicine at Cornell University was also very helpful in the preparation of the manuscript.

## REFERENCES

Bell, E., Ivarsson, B., and Merrill, C. (1979). Production of a tissue-like structure by contraction of collagen lattices by human fibroblasts of different proliferative potential in vitro. Proc. Natl. Acad. Sci. USA 76:1274-1278.

Bellows, C., Melcher, A., and Aubin, J. (1981). Contraction and organization of collagen gels by cells cultured from periodontal ligament, gingiva and bone suggest functional difference between cell types. J. Cell Sci. 50:299-314.

Briggaman, R., Dalldorf, F., and Wheeler, C., Jr. (1971). Formation and origin of basal lamina and anchoring fibrils in adult human skin. J. Cell Biol. 51:384-395.

Carlson, F., Brendel, K., Hjelle, J., and Meezan, E. (1978). Ultrastructural and biochemical analysis of isolated basement membranes from kidney glomeruli and tubules and brain and retinal microvessels. J. Ultrastruct. Res. 62:26-53.

David, G. and Bernfield, M. (1981). Type I collagen reduces the degradation of basal lamina proteoglycan by mammary epithelial cells. J. Cell Biol. 91:281-286.

Emerman, J. and Pitelka, D. (1977). Maintenance and induction of morphological differentiation in dissociated mammary epithelium on floating collagen membranes. In Vitro 13:316-328.

Epstein, E., Jr. (1974). $[\alpha1(III)]_3$ Human skin collagen, release by pepsin digestion and preponderance in fetal life. J. Biol. Chem. 249:3225-3231.

Fessler, L., Timpl, R., and Fessler, J. (1981). Assembly and processing of procollagen type III in chick embryo blood vessels. J. Biol. Chem. 256:2531-2537.

Gay, S., Vilganto, J., Pennttinen, R., and Rackallio, J. (1978). Collagen types in early phases of wound healing in children. Acta Chir. Scand. 144:205-211.

Gross, J. and Kirk, D. (1958). The heat precipitation of collagen from neutral salt solutions: some rate regulating factors. J. Biol. Chem. 233:355-360.

Harris, A., Stopak, D., and Wild, P. (1981). Fibroblast traction as a mechanism for collagen morphogenesis. Nature 290:249-251.

Hassell, J., Robey, P., Barrach, H., Wilczek, J., Rennard, S., and Martin, G. (1980). Isolation of a heparan sulfate containing proteoglycan from basement membrane. Proc. Natl. Acad. Sci. USA 77:4494-4498.

Hayashi, T. and Nagai, Y. (1979). Separation of the $\alpha$ chains of type I and III collagens by SDS-polyacrylamide gel electrophoresis. J. Biochem. 86:453-459.

Helseth, D., Jr. and Veis, A. (1981). Collagen self-assembly in vitro: differentiating specific telopeptide-dependent interactions using selective enzyme modification and the addition of free amino telopeptide. J. Biol. Chem. 256:7118-7128.

Jepsen, A. (1974). An in vitro model of an oral keratinizing squamous epithelium. Scand. J. Dent. Res. 82:144-146.

Jepsen, A., MacCallum, D., and Lillie, J. (1980). Fine structure of subcultivated stratified squamous epithelium. Exp. Cell Res. 125:141-152.

Junqueira, L., Toledo, D., and Montes, G. (1981). Correlation of specific sulfated glycosaminoglycans with collagen types I, II, and III. Cell Tissue Res. 217:171-175.

Kang, A., Nagai, Y., Piez, K., and Gross, J. (1966). Studies on the structure of collagen utilizing a collagenolytic enzyme from tadpole. Biochem. 5:509-515.

Kanwar, Y. and Farquhar, M. (1979). Isolation of glycosaminoglycans (heparan sulfate) from glomerular basement membranes. Proc. Natl. Acad. Sci. USA 76:4493-4497.

Karasek, M. and Charlton, M. (1971). Growth of post-embryonic skin epithelial cells on collagen gels. J. Invest. Derm. 56:205-210.

Klebe, R. (1974). Isolation of a collagen-dependent cell attachment factor. Nature 250:248-251.

Konigsberg, I. and Hauschka, S. (1965). Cell and tissue interactions in the reproduction of cell type. In Reproduction: Molecular, Subcellular and Cellular (M. Locke, ed.), Academic Press, New York.

Kosher, R. and Church, R. (1975). Stimulation of in vitro somite chondrogenesis by procollagen and collagen. Nature (Lond.) 258: 327-330.

Laemmli, U. (1979). Cleavage of structural proteins during the assembly of the head of bacteriophage T4. Nature (Lond.) 227: 680-685.

Lapiere, C., Nusgens, B., and Pierard, G. (1977). Interactions between collagen type I and type III in conditioning bundles organization. Connect. Tissue Res. 5:21-29.

Lenaers, A. and Lapiere, C. (1975). Type III procollagen and collagen in skin. Biochem. Biophys. Acta 400:121-131.

Lillie, J. and MacCallum, D. (1977). Culture of bovine corneal endothelial cells: an in vitro model of basal lamina biosynthesis. J. Cell Biol. 75:161a.

Lillie, J., MacCallum, D., and Jepsen, A. (1978). Growth of subcultivated stratified squamous epithelium on collagen gel rafts. Anat. Record 190:460.

Lillie, J., MacCallum, D., and Jepsen, A. (1980). Fine structure of stratified squamous epithelium subcultivated on collagen rafts. Exp. Cell Res. 125:153-165.

Lillie, J., MacCallum, D., and Jepsen, A. (1981). Extra-cellular matrices in the culture of stratified squamous epithelium. In Vitro 17:215.

MacCallum, D., Lillie, J., Scaletta, L., Occhino, J., Frederick, W., and Ledbetter, S. (1982). Bovine corneal endothelium in vitro: elaboration and organization of a basement membrane. Exp. Cell Res. 139:1-13.

Meier, S. and Hay, E. (1973). Synthesis of sulfated glycosaminoglycans by embryonic epithelium. Dev. Biol. 35:318-331.

Meier, S. and Hay, E. (1974). Control of corneal differentiation by extracellular materials. Collagen as a promoter and stabilizer of epithelial stromal production. Dev. Biol. 38:249-270.

Miller, E., Epstein, E., and Piez, K. (1971). Identification of three genetically distinct collagens by cyanogen bromide cleavage of insoluble human skin and cartilage collagen. Biochem. Biophys. Res. Commun. 42:1024-1029.

Miller, E. (1976). Biochemical characteristics and biological significance of the genetically-distinct collagens. Molecul. and Cellular Biochem. 13:165-192.

Minor, R. (1980). Collagen metabolism: A comparison of diseases of collagen and diseases affecting collagen. Am. J. Path. 98: 227-280.

Patterson, D. and Minor, R. (1977). Hereditary fragility and hyperextensibility of the skin of cats: A defect in collagen fibrillogenesis. Lab. Invest. 37:170-179.

Reddi, H. and Anderson, W. (1976). Collagenous bone matrix-induced endochondral ossification and hemopoiesis. J. Cell Biol. 69:557-572.

Reid, L. and Rojkind, M. (1979). New techniques for cultivating differentiated cells: reconstituted basement membrane rafts. Methods Enzymol. 58:263-278.

Rheinwald, J. and Green, H. (1975). Serial cultivation of strains of human epidermal keratinocytes: the formation of keratinizing colonies from single cells. Cell 6:331-344.

Rhodes, R. and Miller, E. (1978). Physio-chemical characterization and molecular organization of collagen A and B chains. Biochem. 17:3442-3448.

Schor, S. and Court, J. (1979). Different mechanisms in the attachment of cells to native and denatured collagen. J. Cell Sci. 38: 267-281.

Stopak, D. and Harris, A. (1980). Structuring of collagen matrices by cell traction. J. Cell Biol. 87:117a.

Sugrue, S. and Hay, E. (1981). Response of basal epithelial cell surface and cytoskeleton to solubilized extracellular matrix molecules. J. Cell Biol. 91:45-54.

Terranova, V., Rohrbach, D., and Martin. G. (1980). Role of laminin in the attachment of PAM 212 (epithel) cells to basement membrane collagen. Cell 22:719-726.

Timpl, R., Glanville, R., Nowack, H., Wiedemann, H., Fietzek, P., and Kuhn, K. (1975). Isolation, chemical and electron microscopical characterization of neutral-salt-soluble type III collagen and procollagen from fetal bovine skin. Hoppe-Seyler's Z. Physiol. Chem. 356:1783-1792.

Wahl, L., Wahl, S., Mergenhagen, S., and Martin, G. (1974). Collagenase production by endotoxin-activated macrophages. Proc. Natl. Acad. Sci. USA 71:3598-3601.

Wicha, M., Liotta, L., Garbisa, S., and Kidwell, W. (1979). Basement membrane collagen requirements for attachment and growth of mammary epithelium. Exp. Cell Res. 124:181-190.

# PART TWO

# 6

# THE ROLE OF EPITHELIAL-MESENCHYMAL INTERACTIONS IN REGULATING GENE EXPRESSION DURING AVIAN SCALE MORPHOGENESIS

*Roger H. Sawyer*

The skin of vertebrates is derived from the ectodermal epithelium covering the body and, in most cases, the underlying mesenchyme of dermatome or somatopleural origin. Although it is known that interactions between the ectoderm and mesoderm are necessary for the morphogenesis and pattern formation of the skin and its appendages (hairs, feathers, scales, etc.), the mechanisms by which these interactions occur still remain obscure (Brotman, 1977a; 1977b; Dhouailly, 1975; Goetinck, 1980; Kollar, 1981; Lash and Burger, 1977; Sawyer, 1979; Saxen and Karkinen-Jääskeläinen, 1981; Sengel, 1976; Slavkin et al., 1977; Wessells, 1977). Wessells (1977) points out that both instructive and permissive interactions occur in the development of vertebrate skin. However, Wessells cautions his readers that, although "the remarkable effects of dermis on actual feather structure imply that a kind of 'instruction' is affecting the expressive process or morphogenesis," a direct test of the restricted state of the epidermal cells has not been done (Wessells, 1977, p. 59). One possible approach would be to analyze the keratin proteins produced by epidermal cells when associated with different dermises. This would demonstrate that not only are the complex structures of scales and feathers controlled by the dermis, but also specific gene products, the scale- and feather-specific keratins (Kemp, 1975; Kemp and Rogers, 1972).

My intent here is to summarize our knowledge to date on the role of tissue interactions during the morphogenesis and biochemical differentiation of the avian scutate scale. To accomplish this I will first provide a description of normal scutate scale development as well as the abnormal morphogenesis that occurs in the scaleless (sc/sc) mutant chicken, which lacks scutate scales. Reticulate scales which differ not only in their morphogenesis from scutate scales but also in their keratins, will also be discussed as will the spur. Then I will present studies dealing with the epidermal–dermal interactions which occur during normal and abnormal (scaleless) morphogenesis of the integument, including the use of the extraembryonic ectoderm (chorionic epithelium) as a responding tissue in recombination experiments. These experiments demonstrate that the epidermal–dermal interactions involved in morphogenesis of scutate scales are reciprocal and change as development progresses. In addition to the tissue recombination studies, I will show how the biochemical analysis of the major skin proteins (keratins), as well as their immunofluorescent localization using specific antibodies, has allowed us to examine more specifically the regional control of gene expression in the skin as well as the relationship between morphogenesis and biochemical differentiation. These studies will lend support to the conclusion that the dermis does indeed regulate the expression of scale specific keratins by the overlying epidermal cells. This ability of the scutate scale dermis develops through interactions with the overlying epidermis at earlier stages of development. I will show how the results of these foregoing experiments led us to carry out specific epidermal–dermal recombinations which corrected the phenotype of the scaleless epidermis. And finally, I will introduce some studies which are aimed at determining the primary action of the scaleless gene.

## SCUTATE SCALE DEVELOPMENT

The anterior metatarsus of chickens elaborates two rows of scutate scales (Fig. 6.1) which first appear as epidermal placodes at $9\frac{1}{2}$ to 10 days of incubation (Sawyer, 1972a) (a, Fig. 6.1). As seen in feather development (Wessells, 1965), these scale placodes are discrete populations of cells that undergo a period (20–35 hours) during which they cease DNA synthesis (Sawyer, 1972b; Yohro, 1969). During day 11, the scale primordium becomes elevated (b, Fig. 6.1), and by day 12 a definitive scale ridge is formed (c, Fig. 6.1). From 13 to 17 days of incubation the scale ridge elongates distally—so that the scutate scales overlap (d, Fig. 6.1). It is during this time that the outer epidermal surfaces of the scales become distinct, histologically, from the inner epidermal surfaces and hinge regions (Sawyer, 1972a; 1972b). From day 17 to hatching, ultrastructural and biochemical changes occur in the outer and inner epidermal surfaces of the scale so that the adult scale structure is established with a hard plate–like outer scale surface characterized by a Beta Stratum (beta keratins) and a softer more pliable inner scale surface and hinge region with Alpha Strata (alpha keratins) (Baden and Maderson,

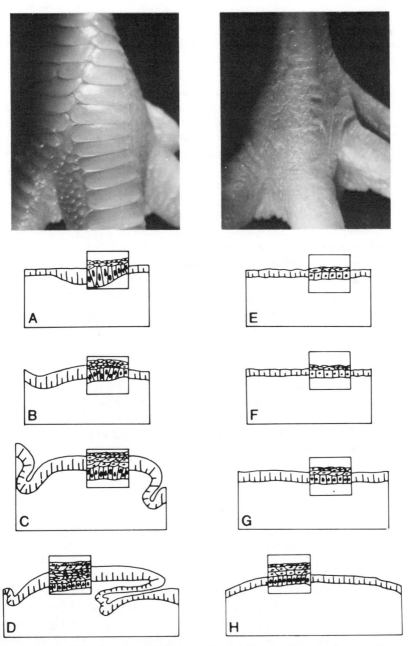

Figure 6.1 This composite figure shows the scutate scales on the foot of a normal hatchling (at the left) and the surface of the foot of a scaleless hatchling (at the right) which lacks scutate scales. The line drawings represent stages in the morphogenesis of normal (A-D) and scaleless (E-H) scutate scale regions at 10 (A and E), 11 (B and F), 12 (C and G), and 16 (D and H) days of development.

1970; Lucas and Stettenheim, 1972; Parakkal and Alexander, 1972; Sawyer et al., 1974a; Sawyer and Borg, 1979; 1980).

It is the plate-like outer epidermal surface of the scutate scale, with its scale specific beta keratins (Dhouailly et al., 1978; O'Guin and Sawyer, 1982), which distinguishes the scutate scale from: (1) the surrounding skin epidermis which produces only alpha keratins; (2) the reticulate scales which produce only alpha keratins; and (3) the scaleless anterior metatarsal skin which elaborates only alpha keratins (O'Guin and Sawyer, 1982). In other words, these beta keratinizing plates develop in a specific pattern, surrounded by more flexible alpha keratinizing skin—i.e., the inner scale surface and hinge region (Sawyer and Borg, 1980).

Unlike scutate scales, the reticulate scales on the plantar surface of the foot are nonoverlapping, radially symmetrical structures (Sawyer and Craig, 1977), which elaborate only one epidermal surface which produces alpha keratins (O'Guin and Sawyer, 1982; Sawyer and Borg, 1979; Spearman, 1966; 1973). They also differ developmentally from scutate scales, in that they do not undergo formation of an epidermal placode. Reticulate scales arise as radially symmetrical anlagen very similar to the formation of scales in snakes and lizards as described by Maderson (1965; 1972). Interestingly, an epidermal placode stage is not found in the embryogenesis of the overlapping scales of alligators which are very similar to the avian scuta (Maderson and Sawyer, 1979).

A recent study of the avian spur demonstrates that although its morphological development is quite similar to that of reticulate scales (Smoak, 1980), the keratins of the spur's epidermis are identical to those found in scutate scale epidermis (Smoak and Sawyer, 1982).

The Scaleless (sc/sc) Mutant Chicken

Chickens homozygous for the autosomal, recessive gene, scaleless (Abbott and Asmundson, 1957), lack completely the scutate and scutellate scales (Fig. 6.1). They also lack most of their feathers and display abnormal morphogenesis of their reticulate scales, spurs, conjunctiva papillae, and scleral ossicles (Blanck et al., 1981; Brotman, 1977a; 1977b; Goetinck, 1980; Goetinck and Sekellick, 1970; 1972; Palmoski and Goetinck, 1970; Sawyer, 1975; 1979; Sawyer and Abbott, 1972; Sawyer and Borg, 1980).

Histological studies (Sawyer and Abbott, 1972) have demonstrated that the initial morphological step in scutate scale development (i.e., epidermal placode formation) does not occur in scaleless embryos (e, Fig. 6.1). Consequently, morphogenesis of a definitive

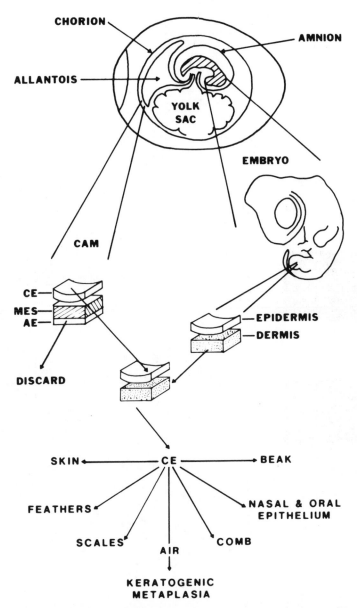

Figure 6.2 This drawing illustrates the origin of the chorioallantoic membrane (CAM) from which the chorionic epithelium (CE) is obtained. After separating the epidermis and dermis of the presumptive integument from numerous regions of the embryo's body, the dermal component is wrapped in CE which then undergoes differentiation into the appropriate epidermal structure.

TABLE 6.1

Response of Chorionic Epithelium (CE) to Various Dermal Tissues

| Responding Tissue | | Source of Dermis | Age | Response | Reference |
|---|---|---|---|---|---|
| 9 day CE | + | Chick Backskin | $6\frac{1}{2}$–8 day | Feather Epidermis | (1) |
| 9 day CE | + | Chick Headskin | 6 day | Feather Epidermis | (2) |
| 9 day CE | + | Chick Beak | 6 day | Beak Epidermis | (2) |
| 9 day CE | + | Chick Comb | 6 day | Comb Epidermis | (2) |
| 9 day CE | + | Chick Nasal Cavity | 6 day | Nasal Lining | (2) |
| 9 day CE | + | Chick Oral Cavity | 6 day | Oral Lining | (2) |
| 9 day CE | + | Chick Metatarsal Skin | 16 day | Scutate Scale Epidermis | (3, 4, 5) |
| 9 day CE | + | Chick Footpad Skin | 16 day | Reticulate Scale Epidermis | (5) |
| 9 day CE | + | Chick Scaleless Metatarsal Skin | 16 day | Scaleless Metatarsal Epidermis | (4, 5) |
| 9 day CE | + | Chick Scaleless Footpad Skin | 16 day | Scaleless Footpad Epidermis | (5) |
| 9 day CE | + | Human Palm or Sole Skin | Fetal | Chick Epidermis | (6) |
| 9 day CE | + | Monkey Sole or Ear Skin | Fetal | Chick Epidermis | (7) |
| 9 day CE | + | Lizard Tail Skin | Embryonic | Chick Epidermis | (8) |

Sources: (1) Watson (1968); (2) Kato and Hayashi (1963); (3) Kato (1969); (4) Sawyer and Abbott (1972); (5) Sawyer (1979); (6) Sawyer, Abbott and Treford (1972); (7) Sawyer (1975); (8) Sawyer (unpublished data).

120

scale ridge and its outgrowth do not occur and the anterior metatarsus of the scaleless embryo remains flat (e-h, Fig. 6.1). Histological (Sawyer and Abbott, 1972), ultrastructural (Sawyer et al., 1974b), biochemical (O'Guin and Sawyer, 1982) and X-ray diffraction (Baden et al., 1975) studies demonstrate that the scaleless anterior metatarsal skin never develops the Beta Stratum which characterizes the outer epidermal surface of scutate scales, only an Alpha Stratum develops (Sawyer, 1979).

In contrast, recent studies show that the reticulate scales do appear in scaleless, although they are very irregular in shape (Sawyer and Craig, 1977; Sawyer, 1979). The soluble proteins of their epidermis (Mirkes and Sawyer, 1979) as well as their pattern of keratinization are identical to normal reticulate scales (Sawyer and Borg, 1979; 1980; O'Guin and Sawyer, 1982).

The spur, on the other hand, undergoes both abnormal morphogenesis and abnormal keratin production (Smoak and Sawyer, 1982) in the scaleless mutant.

The Chorionic Epithelium (CE) as a Responding Tissue

Moscona (1959; 1960) suggested that the chick chorionic epithelium (CE) might prove useful as a tissue to study embryonic differentiation. Kato and Hayashi (1963) and Kato (1969) further demonstrated the usefulness of the CE as a responding tissue, in an elegant set of experiments, in which the CE was induced to form normal epidermal structures under the influence of several embryonic dermises. Figure 6.2 illustrates the approach taken when combining the CE with mesodermal tissues of the chick embryo.

The CE responds to the dermis from specific regions by forming the appropriate epidermis. The age of the dermis is important, as will be discussed later. In combination with dermal tissues from different species, the CE responds by forming a generalized chick epidermis (a summary of the response of the CE to dermal tissue is given in Table 6.1). This supports the conclusions of Dhouailly et al. (1978), that in the epithelial-mesenchymal interactions between tissues of different species the epithelium can only respond within the range of its own genome.

EPIDERMAL-DERMAL INTERACTIONS
OF AVIAN INTEGUMENT

Reciprocal heterotypic recombinations between the epidermal and dermal components of feather- and scutate scale-forming regions

(Rawles, 1963; Sengel, 1958) have demonstrated that the 9-13-day scale epidermis forms only feathers when combined with 5-8.5-day feather dermis. Although Rawles (1963) does not present the histology, it is assumed that these feathers developed de novo; that is, from epidermal placodes. Aberrant feathers also form in these recombinants when the scale epidermis is 12 days old or older. In combination with feather epidermis (5-8.5 days), the scale dermis undergoes a shift from initially supporting feather formation (9-12 days) to eventually supporting scale formation when 13, 14, or 15 days of age. Rawles (1963) concluded that the anterior metatarsal dermis was bipotential in nature, first eliciting feather formation and then scutate scale formation from presumptive feather epidermis.

Kato's (1969) work demonstrated that the CE forms perfectly normal scutate scale epidermis (both outer and inner surfaces) when carefully combined with the outer and inner surfaces of the dermal ridges of 15-day scutate scales. Keep in mind that the scale ridges are already present at this stage of development. New scales do not form de novo from new epidermal placodes. As pointed out by Sawyer and Abbott (1972, pp. 107-108), when the CE is combined with 16-day

TABLE 6.2

Response of Chorionic Epithelium (CE) to Anterior Metatarsal (AM) Dermis of Differing Embryonic Ages

| Responding Tissue | | Source of Dermis | Age | Response | Reference |
|---|---|---|---|---|---|
| 8 or 9 day CE | + | Chick (AM) | < 10 day | Feathers | (1) |
| 8 or 9 day CE | + | Chick (AM) | 11 day | Barb Ridges | (2) |
| 8 or 9 day CE | + | Chick (AM) | 12 day | Barb Ridges & Scutate Scales | (2) |
| 8 or 9 day CE | + | Chick (AM) | 13 day | Scutate Scales | (2, 3) |
| 8 or 9 day CE | + | Chick (AM) | 14 day | Scutate Scales | (3) |
| 8 or 9 day CE | + | Chick (AM) | 15 day | Scutate Scales | (4) |
| 8 or 9 day CE | + | Chick (AM) | 16 day | Scutate Scales | (3) |

Sources:
  (1) Kawaga and Kato (personal communication)
  (2) Fisher and Sawyer (1979)
  (3) Sawyer (1979)
  (4) Kato (1969)

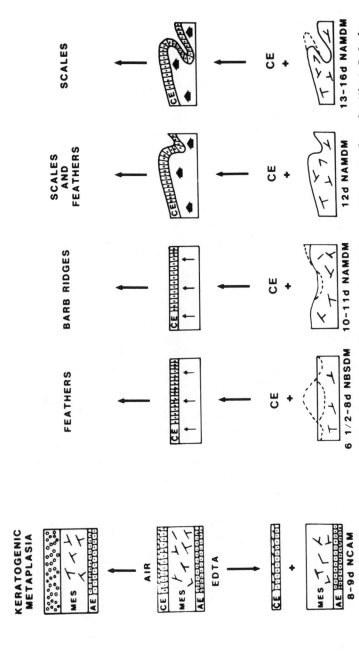

Figure 6.3 This diagram illustrates the results obtained when 8–9-day CE is combined with 6.5–8-day normal backskin dermis (NBSDM), 10–11-day normal anterior metatarsal dermis (NAMDM), 12-day normal anterior metatarsal dermis (NAMDM), or 13–16-day normal anterior metatarsal dermis (NAMDM). When exposed to air the CE undergoes keratogenic metaplasia.

123

anterior metatarsal dermis, it "adheres to the naked dermal ridges and then undergoes transformation into outer and inner epidermal surfaces. The flat CE basal cells are transformed into oriented columnar basal cells (Kato, 1969). Typical scale placodes do not form. Thus placode formation is not essential to histogenesis of scale epidermis once the dermal ridge is formed."

Next we set out to determine if the highly plastic CE would also respond to the temporal changes that occur in the normal anterior metatarsal dermis as proposed by Rawles (1963). We found that indeed the anterior shank dermis changes as it ages from 11 to 13 days of incubation (Fisher and Sawyer, 1979). Only barb ridges formed when CE was combined with 11-day anterior metatarsal dermis while both barb ridges and scutate scales formed with 12-day dermis and only scutate scales with 13-day dermis (Fig. 6.3, Table 6.2). Previously, Kawaga and Kato (personal communication) had found that only feathers develop when the CE was recombined with anterior metatarsal dermis, less than 10 days of age, and Dhouailly (personal communication) found that the anterior metatarsal dermis from 10-day quail embryos will elicit feather formation when grafted under the ectoderm of the 2-day somatopleura of chick embryos. That the anterior metatarsal dermis, 13 days or older, is capable of inducing scutate scale epidermis from several responding epithelia is well documented. Backskin epidermis, corneal epithelium, beak epidermis, and apteric epidermis, as well as the chorionic epithelium will form scutate scale epidermis with scutate scale dermis (Table 6.3).

Unlike the results obtained with the CE, the anterior metatarsal dermis can, in fact, direct the formation of scutate scales at early stages of development with certain epidermises (Table 6.4), as demonstrated by Linsenmayer (1972) and Dhouailly and Sengel (this volume). Nine-day anterior metatarsal dermis, just prior to the time at which epidermal placodes form, will induce scutate scale formation in 11-day footpad (reticulate scale region) epidermis (Linsenmayer, 1972). Likewise, either 8.5-day or 10-day anterior metatarsal dermis will elicit scutate scale formation in combination with 10-day apteric epidermis (Dhouailly and Sengel, this volume). These results are inconsistent with those obtained with the chorionic epithelium, however, as pointed out earlier (Sawyer, 1979, p. 12), "Although the CE and young epidermis are similar in some respects (Kato, 1969; Sawyer, 1978), they apparently differ in their morphogenetic potential. The CE cells are able to become epidermal cells under the direction of the anterior shank dermis, yet they do not possess the ability to form the epidermal placodes of scutate scales, since they are not 'leg or foot' epidermal cells."

It is not appropriate to extend the lack of scale placode formation by the CE to all situations since normal feathers readily form

TABLE 6.3

Summary of Heterotypic Tissue Recombinations in Which
Anterior Metatarsal (AM) Dermis (Older Than 13 Days)
Is Used as the Inductive Tissue

| Inductive Tissue | Age | | Responding Tissue | Age | Response | References |
|---|---|---|---|---|---|---|
| Chick AM Dermis | (> 13 d) | + | Chorionic Epithelium | (8-9d) | Scutate Scales | (1, 2) |
| Chick AM Dermis | (> 13 d) | + | Backskin | (5-8½d) | Scutate Scales and Feathers | (3) |
| Chick AM Dermis | (13 d) | + | Corneal Epithelium | (5d) | Scutate Scales | (4) |
| Chick AM Dermis | (13 d) | + | Chick Beak Epidermis | (5-7½d) | Scutate Scales | (3) |
| Chick AM Dermis | (13 d) | + | Apteric Epidermis | (5-7½d) | Scutate Scales | (3) |

Sources:
  (1) Kato (1969)
  (2) Sawyer (1979)
  (3) Rawles (1963)
  (4) Coulombre and Coulombre (1971)

when the CE is combined with presumptive feather dermis; either
6-day head dermis (Kato and Hayashi, 1963) or 6.5-8-day back skin
dermis (Watson, 1968). That the CE will readily form feathers de
novo with presumptive feather dermis, yet will not form scutate
scales de novo with presumptive scuta dermis, suggests that there
is either a real difference in the initial inductive mechanisms for
feathers and scutate scales and/or that the CE has a feather bias.
    Does the CE possess an intrinsic ability to form feathers?
Dhouailly (1978) confronted the extraembryonic somatopleura of 2-3-
day chick embryos with a nonspecific stimulus, 12.5-13-day embryonic mouse lip dermis. In the ectoderm overlying the mouse dermal
cells, feather buds with aberrant barb-ridges formed, while normal
feathers formed near the mouse dermal cells from the extraembryonic ectoderm and its own mesenchyme. Dhouailly (1978) proposes that
by causing the host mesenchyme of the somatopleura to transform into
a "dense dermis," feather formation is achieved. Dhouailly and Sengel
(this volume) conclude that "early in development, probably at the time

TABLE 6.4

Summary of Heterotypic Tissue Recombinations in Which
Anterior Metatarsal (AM) Dermis (Younger Than 13 Days)
Is Used as the Inductive Tissue

| Inductive Tissue | Age | Responding Tissue | Age | Response | Reference |
|---|---|---|---|---|---|
| Chick AM Dermis | (< 12d) + | Chorionic Epithelium | (8-9d) | Feathered Epidermis | (1) |
| Chick AM Dermis | (9d) + | Footpad Epidermis | (11d) | Scutate Scales | (2) |
| Chick AM Dermis | ($8\frac{1}{2}$ or 10d) + | Apteric Epidermis | (10d) | Scutate Scales | (3) |
| Chick AM Dermis | (10-11d) + | Scaleless AM Epidermis | (10-11d) | Scaleless Skin | (4) |
| Chick AM Dermis | (11-12d) + | Scaleless AM Epidermis | (14-15d) | Scaleless Skin | (5) |
| Chick Leg Bud Mesoderm | (3d) + | Scaleless Leg Bud Ectoderm | (3d) | Scaleless Limb | (6) |

Sources:
  (1) Kawaga and Kato (personal communication); Fisher and
      Sawyer (1979)
  (2) Linsenmayer (1972)
  (3) Dhouailly and Sengel (this volume)
  (4) Sengel and Abbott (1963)
  (5) McAleese and Sawyer (1981)
  (6) Goetinck and Sekellick (1972)

of or even before gastrulation, presumptive ectoderm becomes subdivided into two categories; embryonic ectoderm (later epidermis),
with full appendage-forming competence; and extraembryonic ectoderm (later amniotic and chorionic epithelium) with a diminished
competence, restricted to feather placode formation" (Dhouailly and
Sengel, this volume).

   If one considers the response of the CE to various mesenchymal
tissues (Table 6.1, Fig. 6.2), it can be concluded that the CE makes
the appropriate epidermis in response to a mesenchyme (dermis)
only after the basic form of the structure (i.e., beak, comb, scuta,
or reticula) is present. The CE is capable of becoming an epidermis
which synthesizes the appropriate products in response to a specific

dermis. The exception is in the case of feather formation where feather placodes form.

One other point concerning the extraembryonic epithelia as responding tissues: Kollar (this volume) has stressed that very early in development the endodermal and ectodermal and probably mesodermal tissues become restricted in their abilities to express certain phenotypes. Some years ago I made at least 100 recombinants between 15-day scutate scale dermis and the endodermal epithelium from the allantois of the 8-day chorioallantoic membrane. Although some stratified squamous epithelium was occasionally obtained, scutate scale epidermis never formed. Apparently, the separation of the embryo into ectoderm, endoderm, and mesoderm is not done just as a convenience to students of embryology but is an actual functional separation of the embryonic tissues, and may relate to the regulation of gene expression.

EPIDERMAL-DERMAL INTERACTION
WITH SCALELESS

Epidermal-dermal recombinations between normal and scaleless skin at early stages of development (Table 6.4) have demonstrated that the scaleless defect is expressed by the epidermal component, while the dermal component functions normally. Goetinck and Abbott (1963) and Goetinck and Sekellick (1970; 1972) tested recombinations using limb bud tissues, while Sengel and Abbott (1963) tested both backskin and anterior metatarsal skin (Table 6.4) at early stages of feather or scale development (Fig. 6.4).

How does the CE respond to the 15- or 16-day scaleless anterior metatarsal dermis, which lacks ridges? We considered the possibilities that either the scaleless dermis would retain its ability to function normally (Goetinck and Abbott, 1963; Goetinck and Sekellick, 1972; Sengel and Abbott, 1963), and the CE would somehow undergo scale ridge formation to form normal scutate scales, even though the dermis was 5 to 6 days beyond the time of normal placode formation; or that the scaleless dermis, being without the dermal ridges characteristic of the scutate scales, would induce only one of the epidermal surfaces of scutate scales—either the outer epidermal surface with its Beta Stratum or the inner epidermal surface with its Alpha Stratum. The results are clear. The older scaleless dermis (anterior metatarsus) cannot undergo scale ridge formation nor can it direct the formation of the outer scale surface with its scale specific beta keratins, when combined with CE (Sawyer and Abbott, 1972; Sawyer, 1975; Sawyer, 1979). The scaleless dermis directs the CE to form an epidermis identical to that normally seen on the scaleless anterior metatarsus (Sawyer et al., 1974b; Sawyer, 1975), which is very similar to the epidermis along the inner surface and hinge region of the scutate scale. I have proposed that the scaleless anterior metatarsal dermis becomes defective in its inductive abilities because it has not received the appropriate cues from a normal scutate scale epidermis as it undergoes morphogenesis (Sawyer, 1979). In other words, as the normal epidermis and dermis undergo morphogenesis they progressively, and together, produce a three-dimensionally arranged dermal ridge which will provide instructional information for region-specific production of keratins.

Since the scaleless anterior metatarsal dermis starts out functioning normally, when does it become defective? The first consideration was to use the CE as a responding tissue to test the scaleless dermis from 9 to 15 days of incubation, however we now know that the CE cannot make the epidermal placodes of scutate scales— only feathers form. The neutral tissue which should provide the answer to this question is the apteric epidermis, which Dhouailly and Sengel (this volume) have shown is one of the few epidermises that can make scutate scale epidermis with presumptive scutate scale dermis (the other is reticulate scale epidermis) (Linsenmayer, 1972). Once the time of change in the dermis is determined we can then examine the structural and biochemical alterations which occur in the scaleless dermis when it loses its ability to induce the outer epidermal surfaces of scutate scales.

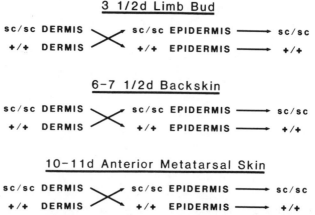

Figure 6.4 This figure illustrates the resulting phenotype in reciprocal tissue recombinations between the epidermis and dermis of normal (+/+) and scaleless (sc/sc) presumptive skin from 3.5-day limb buds, 6-7.5-day backskin or 10-11-day anterior metatarsal skin.

That some changes do occur in scaleless anterior metatarsal dermis is suggested by Brotman (1977b). In recombinant grafts between 7-day feather epidermis and 9- to 18-day scaleless anterior metatarsal dermis Brotman (1977b) found that unlike normal anterior metatarsal dermis which ceases to support feather formation when it reaches 13-14 days of development, the scaleless anterior metatarsal dermis continues to support a low level of feather formation until it reaches 16 days of development. However, he also noticed that at 12 to 13 days of incubation both the normal and scaleless dermis begin to lose their ability to support feather development.

What about reticulate scales? Sawyer and Borg (1979) have shown that although the reticulate scale epidermis gives an alpha-type X-ray diffraction pattern, as does the inner epidermal surface and hinge region of scutate scales, the fine structural features are very different for these two epidermises. When 16-day normal foot-pad (reticulate scale region) dermis is combined with 8-9-day CE (Table 6.1), the CE forms normal reticulate scale epidermis with the alpha keratin which is characteristic of reticulate scale epidermis (Sawyer, 1979; Sawyer and Borg, 1979). The normal form of the reticulate scales is present at 16 days, therefore, the CE is induced to become a reticulate scale epidermis; de novo formation of reticulate scales does not occur. When combined with the 16-day scaleless footpad dermis, the CE responds by making scaleless reticulate scale epidermis; i.e., the shape of the reticulate scale is abnormal (very irregular) yet the alpha keratin which characterizes the normal reticulate scale is present (Sawyer, 1979; Sawyer and Borg, 1980). Recall that an epidermal placode does not form in the development of reticulate scales (Sawyer and Craig, 1977; Sawyer and Borg, 1980). We suggest that since the entire surface of the reticulate scale is only one type of keratin; unlike the scutate scales which form rectangular plates of beta keratin surrounded by regions of alpha keratin, there is no need to establish regional differences in the inductive properties of the reticulate scale dermis (Sawyer and Borg, 1980). Therefore, the abnormal morphogenesis of the reticulate scales in scaleless does not alter the production of keratins. At present we are determining if the CE can respond to the presumptive reticulate scale dermis by making normal reticulate scales de novo, since an epidermal placode stage is not required. Positive results will then allow us to determine the stage in reticulate scale development when the abnormal morphogenesis first becomes established in reticulate scales of scaleless embryos.

The spurs of scaleless are also abnormal morphologically (Abbott and Asmundson, 1957; Smoak, 1980). It has been suggested that these large cornified protuberances (with an ossious core) are modified reticulate scales (Lucas and Stettenheim, 1972) and like the reticulate scales and spurs of scaleless embryos are abnormally shaped (Smoak, 1980). However, Smoak and Sawyer (1982) have shown that the beta keratins found in the normal spur are missing in scaleless as are some of the alpha keratins. This suggests that the

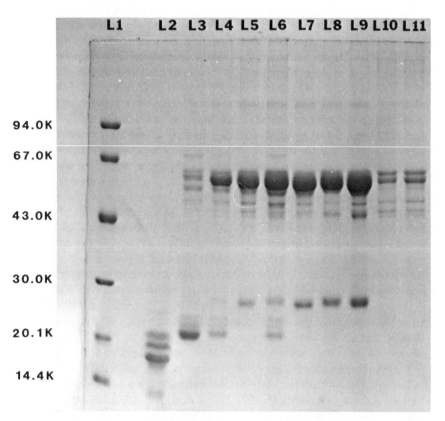

Figure 6.5 This is a photograph of a Coomassie brilliant blue stained polyacrylamide gel which illustrates the various S-carboxymethylated (SCM) keratin polypeptides obtained from different regions of the integument from normal and scaleless chick embryos. Lane 1 (L1) shows molecular weight standards. Lane 2 (L2) shows the SCM derivatives from 20-day embryonic down feather while lane 3 (L3) shows the polypeptides from the outer surface of scutate scales from 12-day hatchlings. Lanes 4-11 (L4-L11) show the polypeptides from the normal anterior metatarsal epidermis (L4), scaleless anterior metatarsal epidermis (L5), normal posterior metatarsal epidermis (L6), scaleless posterior metatarsal epidermis (L7), normal footpad epidermis (L8), scaleless footpad epidermis (L9), normal backskin epidermis (L10), and scaleless backskin epidermis (L11).

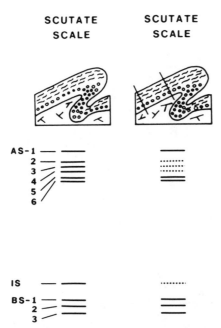

Figure 6.6 This drawing illustrates the gel patterns obtained from whole scutate scale epidermis (left) and the epidermis from the outer surface only (right). Although all bands are found in the epidermis of the outer surfaces AS2-4 and IS-2 are greatly reduced.

scaleless gene also affects the spur in a manner similar to its effect on scutate scales.

Tissue-Specific Keratins as Indicators of Gene
Expression in Avian Skin

Since Lockett et al. (1979) showed that there are at least 56 different genes coding for beta keratin alone, and because there are regions of the avian integument which give the same X-ray diffraction patterns yet differ histologically and fine structurally (Sawyer and Borg, 1980), we set out to identify the major tissue-specific keratin polypeptides actually present in the different regions of the integument (O'Guin and Sawyer, 1982).

With minor modifications of the methods outlined by Dhouailly et al. (1978) we separated the S-carboxymethylated (SCM) polypeptides, which were labeled with $^{14}$C-iodoacetate, by SDS-polyacrylamide gel electrophoresis (O'Guin and Sawyer, 1982). We identify

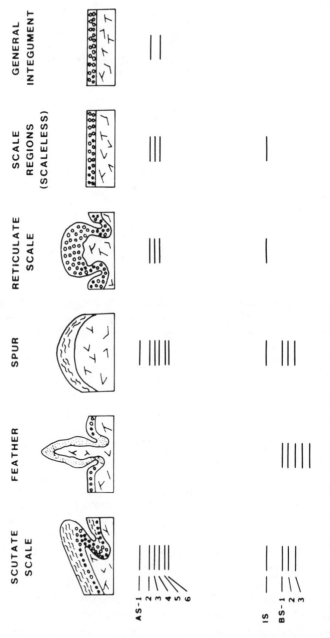

**SDS-PAGE OF KERATIN POLYPEPTIDES**

Figure 6.7 This drawing illustrates the keratin polypeptides obtained from several regions of the integument including all scale regions of the scaleless mutant.

132

ten keratin bands in scutate scale epidermis, which we designate as AS-1 to 6 (alpha scale keratins molecular weight from 55.5 k to 70.0 k), IS-1 (intermediate scale keratin molecular weight of 26.0 k) and BS-1 to 3 (beta scale keratins molecular weight from 19.5 k to 22.6 k). Feathers give five beta keratin bands, two of which are of a lower molecular weight than the scutate scale bands BS-1 to 3. Figure 6.5 shows a typical gel stained with Coomassie brilliant blue. Dhouailly et al. (1978) assumed, as we did, that the alpha keratin bands found in the scutate scale epidermis were due to contamination from the inner epidermal surfaces and hinge regions; however by examining the outer epidermal surface separated from the other regions (Fig. 6.6) we found all six alpha bands still present, yet AS-1, 5, and 6 were enhanced over AS-2, 3, and 4 (O'Guin and Sawyer, 1982). All ten bands are considered to represent the biochemical differentiation of scutate scale epidermis. At present we do not know if every band is a specific gene product or if some post-transcriptional or transla-tional modifications are occurring.

Analysis of the scaleless anterior metatarsal epidermis shows the total absence of BS-1 to 3 as well as the absence of AS-1, 5, and 6 (Fig. 6.7). This suggests that the lack of scutate scale morpho-genesis in scaleless can be correlated with the absence of specific gene products.

Normal reticulate scales elaborate only AS-2, 3, and 4 as well as the IS-1 band, and no beta keratin is produced. An identical pat-tern is found for scaleless reticulate scales (Fig. 6.7), verifying that the abnormal morphogenesis of scaleless reticulate scales does not affect their terminal biochemical differentiation. Comparison of scaleless backskin (general integument) with normal backskin, without its feathers (plucked), shows identical patterns which differ from any of the scale-forming regions of either normal or scaleless (Fig. 6.7). Interestingly, the banding profile for normal spur epidermis, which is supposedly a modified reticulate scale, is identical to scutate scales, not reticulate scales. The scaleless spur which is abnormal in its morphology is also abnormal in its keratin production. We propose that the role of the epidermis in establishing the inductive ability of the spur dermis is defective in scaleless (Smoak and Sawyer, 1982). Thus, the scaleless gene affects various skin derivatives in different ways which we believe relates to differences in the epithelial-mesenchymal interaction of the various derivatives.

To better clarify the specific location of the alpha and beta keratins in the scutate scale epidermis, we produced non-cross-reacting antisera to alpha and beta keratins (O'Guin et al., 1982). Recall that the gel profiles of the keratins from the scutate scale epidermis showed the presence of alpha keratin (AS-1 through 6) in the outer plate-like surface which is characterized by its Beta

Stratum (Sawyer et al., 1974a; Sawyer, 1975; 1979). Using indirect immunofluorescent procedures the antisera clearly show a vertical arrangement of the alpha and beta keratins in the outer epidermal surface of scutate scales. Alpha keratin is present in the stratum basal and stratum intermedium while the beta keratin is found in the stratum intermedium and stratum corneum (O'Guin et al., 1982).

When and where are the messenger RNAs made for these keratins and how are they regulated by epithelial-mesenchymal interactions? To begin to answer these questions I have utilized the dot blot technique (Thomas, 1980) with a plasmid (pCSK-12) containing a beta keratin sequence (a generous gift from Dr. George E. Rogers, University of Adelaide, South Australia) and total RNA (extracted by the procedure of Chirgwin et al., 1979) from the anterior metatarsal skin (scutate scales) of 20-day normal and scaleless embryos. While positive hybridization occurs with the RNA from normal scutate scales, hybridization is not detectable with the RNA from scaleless embryos, indicating that the message for this beta keratin sequence is not present in the scaleless anterior metatarsal skin.

Correcting the Scaleless (sc/sc) Phenotype

The reciprocal recombinations between the epidermal and dermal components of normal and scaleless anterior metatarsal skin (10-11-day) demonstrated that the scaleless epidermis was defective yet the scaleless dermis was capable of forming normal scales when combined with normal epidermis (Sengel and Abbott, 1963). Does the lack of epidermal placodes in the scaleless anterior metatarsal epidermis mean that these epidermal cells will be incapable of elaborating a Beta Stratum? Recombinations of older scaleless dermis with the chorionic epithelium (CE) have shown that the scaleless dermis also becomes defective, i.e., is incapable of eliciting scutate scale formation (neither the scale ridge nor a Beta Stratum form) from the CE (Sawyer, 1975; 1978; 1979). Knowing that the normal scutate scale dermis must be at least 12 days of age in order to direct the formation of scutate scale epidermis from the CE (Fisher and Sawyer, 1979; Sawyer, 1979) or feather epidermis (Rawles, 1963), we reasoned that perhaps the scaleless anterior metatarsal epidermis was not receiving the appropriate cues from the scaleless dermis (McAleese and Sawyer, 1981; 1982). In other words, although the scaleless epidermis cannot form a scale placode, perhaps it can respond to the inductive cues of older scale dermis, which direct the formation of outer and inner scale surfaces with their corresponding Alpha and Beta Strata. Therefore we combined 10-16-day scaleless epidermis with either 14, 15, or 16-day normal scutate

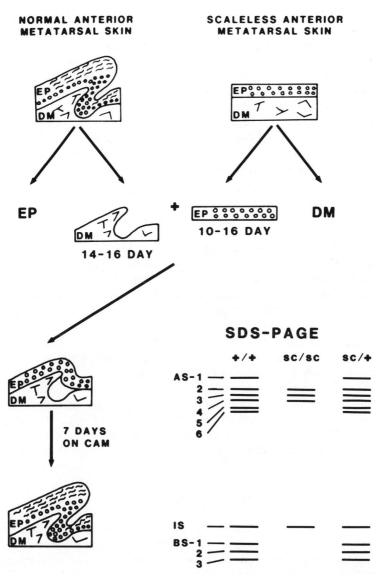

Figure 6.8 This drawing illustrates that normal scutate scales form
when 14-16-day normal anterior metatarsal dermis (DM) is combined
with 10-16-day scaleless anterior metatarsal epidermis (EP) and
grown for 7 days on the chorioallantoic membrane (CAM). Sodium
dodecyl sulfate polyacryamide gel electrophoresis (SDS-PAGE) of the
S-carboxymethylated (SCM) polypeptides of normal (+/+), scaleless
(sc/sc) and the scaleless-normal (sc/+) recombinant shows that the
keratin polypeptides from the recombinant are identical to those from
normal scutate scales.

scale dermis (Fig. 6.8). Normal scutate scales with Beta Strata formed in all cases and the scutate scale specific keratins were all present. These studies support the proposal that the lack of placode morphogenesis by the scaleless epidermis, results in a scaleless dermis which never acquires the ability to induce the outer epidermal surface (Beta Stratum) which characterizes the scutate scale (McAleese and Sawyer, 1981; 1982).

In addition, McAleese and Sawyer (1981; 1982) also examined the inductive ability of the scutate scale dermis from 10 to 16 days of age in combination with 14-15-day scaleless anterior metatarsal epidermis. It was found that the scutate scale dermis younger than 12 days was unable to induce scutate scales, again supporting the contention that morphogenesis of the epidermal placode into a defin-itive scale ridge plays an important role in establishing the inductive ability of the scutate scale dermis.

The recombinants of scaleless epidermis (14-15-day) with 11-day normal scutate scale dermis not only did not form scales, they did not form feathers either. This can be interpreted in two ways. Either the young scutate scale dermis is providing only the instructions for feather placode formation (Rawles, 1963; Fisher and Sawyer, 1979) which are not received or interpreted by the scaleless epidermis, or the young scutate scale dermis is providing instructions for scutate scale placode formation (Dhouailly and Sengel, this vol-ume), to which the scaleless epidermis cannot respond. Brotman (1977a; b) suggests, from epidermal-dermal recombination studies between scaleless high line and normal chick thigh skin, that the presence of feathers on the anterior metatarsis of scaleless high line embryos is due to an increased competence for feather forma-tion by the scaleless epidermis, rather than a change in the dermis.

## What Is the Primary Defect in Scaleless?

In the case of scutate scales, the initial defect in scaleless undoubtedly relates somehow to the lack of epidermal placode forma-tion. Abnormal keratin production in scaleless is a secondary effect, not the primary genetic lesion. That the scaleless genome contains the genes for the scale beta keratins is already known, from three sources. Embryonic scaleless epidermis which makes a subperi-dermal layer (Sawyer et al., 1974b) gives bands BS-1 to 3 (O'Guin and Sawyer, 1982) with gel electrophoresis, scaleless epidermis responds to normal scutate scale dermis with beta keratin production (McAleese and Sawyer, 1982), and X-ray diffraction data demonstrate beta keratins in scaleless claws (Baden et al., 1975).

To approach the question of the scaleless defect we can speak in terms of signals emanating from the tissues. This is a meaningful approach to investigations dealing with epithelial-mesenchymal interactions, and it provides us with a way to define temporal and spatial aspects of the interactions (even though these events are most likely continuous). This then gives us information concerning the most likely stage in development at which we should search for biochemical anomalies in specific tissues. Unfortunately, no one has defined any signal in biochemical terms. If we accept the premise that formation of an epidermal placode is a necessary prerequisite to the formation of a scutate scale ridge and that the establishment of the scutate dermal ridge is required for the eventual expression of the keratins in a specific scutate scale pattern, then a comparison of the biochemical parameters of normal and scaleless tissues at these stages of development is warranted. Since the epidermis and dermis are separated by a basement membrane, it is reasonable to assume that the signaling process occurs at or through this structure. Collagens, glycosaminoglycans, proteoglycans, and the glycoproteins are major components of this extracellular, epidermal-dermal junction (Kollar, this volume; Lillie et al., this volume).

That a functional defect exists in the basement membrane region of scaleless skin has been described (Sawyer and Abbott, 1972; Sawyer, 1979). Normal anterior metatarsal skin requires treatment with a chelating agent or trypsin in order to be separated into its epidermal and dermal components, yet the scaleless anterior metatarsal skin (9-17 days) can be easily separated into epidermis and dermis without chelating agents or enzymes (Sawyer, 1979). This separation of the epidermal and dermal components of scaleless skin occurs between the lamina densa and the underlying dermis (Sawyer, unpublished observations). Are there any biochemical differences in the extracellular components that make up the extracellular matrix of normal and scaleless tissues?

In the case of feather development, Goetinck (1980) has found no differences in the synthesis of collagen between scaleless and normal embryos (at 6 to 11 days of incubation) but finds a defective collagen lattice in the scaleless dermis (Goetinck, 1980; Goetinck and Sekellick, 1972). It has been proposed that this collagen lattice of normal dermis regulates the site at which the feather dermal condensations form (Stuart and Moscona, 1967). Using the electron microscope, Overton and Collins (1976) find that the collagen network of scaleless back skin is more irregular than the normal network and the diameter of the individual fibers is more variable in scaleless.

In the case of sulfated proteoglycans, Goetinck (1980) does find reduced synthesis in scaleless backskins. These molecules bind and stabilize collagen (Toole and Lowther, 1968; Toole et al., 1977).

As far as the glycoproteins are concerned, we have recently examined the distribution of fibronectin (Yamada et al., 1977) in feather morphogenesis using indirect immunofluorescence with an antiserum provided by Dr. Ken Yamada (Haake and Sawyer, 1982). We find that fibronectin is present throughout the normal dermis and that the characteristic anchoring filaments of feather primodia (Wessells, 1965; Kallman et al., 1967; Kischer and Keeter, 1973) are fibronectin positive. Fibronectin is also present in the dermis of scaleless backskin, as detected by indirect immunofluorescence, yet the fibronectin-positive anchoring filaments are more dispersed (Haake and Sawyer, unpublished observations). The proposed interactions between the various components of the extracellular matrix (Engrall et al., 1978; Gold and Pearlstein, 1980; Toole et al., 1977) suggest several possible sites for defects in the scaleless extracellular matrix, yet definite conclusions must await a better understanding of the relationships between the various components of the extracellular matrix and specific morphogenetic events. Since the scaleless epidermis can be easily separated from the underlying dermis without chelating agents or trypsin (Sawyer and Abbott, 1972), I have suggested that some defect exists in those extracellular components which anchor the lamina densa to the underlying dermis (Sawyer, 1979).

It is well known that cell surfaces are intimately involved in adhesion phenomena. Perhaps the mixing of dissociated scaleless skin cells with normal skin cells can correct the adhesion defect (Knapp and Sawyer, 1982a). As an approach to begin to characterize the cell surfaces of the normal and scaleless cells, we have examined the effects of lectins on the agglutination of dissociated epidermis of normal and scaleless 8-day backskin (Knapp and Sawyer, 1982b). In the presence of wheat germ agglutinin (WGA) the scaleless epidermal cells form larger agglutinates in rotation cultures than do normal epidermal cells. Equivalent agglutination is seen for scaleless and normal cells in the presence of concanavalin A, while neither scaleless nor normal epidermal cells agglutinate in the presence of Dolichos biflorus, soybean or Ulex europeus agglutinin. The use of $^3$H-acetyl wheat germ agglutinin demonstrates that there are indeed more WGA binding sites on the scaleless cell surfaces than on the normal cell surfaces (Knapp and Sawyer, 1982). How this difference in lectin binding relates to the scaleless defect is unknown, yet it is intriguing that the initial defect in scaleless is expressed by the epidermis (Sengel and Abbott, 1963; Goetinck and Abbott, 1963).

## CONCLUDING REMARKS

As pointed out in the preface the nature of tissue interactions varies for each system being studied, and in the case of the avian integument; scutate scales, reticulate scales, spurs and feathers all develop differently and the scaleless gene effects each of these skin appendages in a unique manner. It is our hope that continued comparison of the development of normal and scaleless skin appendages at the morphogenetic, biochemical and molecular levels will eventually lead us to an understanding of the mechanisms by which epithelial-mesenchymal interactions act.

## ACKNOWLEDGEMENTS

I wish to thank Peggy Sullivan for her technical assistance and Debra Chavis for typing the manuscript. Also, I am sincerely grateful to the members of my laboratory, Cindy Blanck, Anne Haake, Dr. Loren Knapp, Gerd Koenig, Susan McAleese, and Michael O'Guin, whose diligent work continues to add to our understanding of tissue interactions in development of the integument. This work is supported by National Science Foundation Grant PCM-8011745.

## REFERENCES

Abbott, U. K. and Asmundson, V. S. (1957). Scaleless, an inherited ectodermal defect in the domestic fowl. J. Hered. 18:63-70.

Baden, H. P. and Maderson, P. F. A. (1970). Morphological and biophysical identification of fibrous proteins in the amniote epidermis. J. Exp. Zool. 174:225-232.

Baden, H. P., Lee, L. D., and Kubilus, J. (1975). The structural proteins of scaleless-mutant chick epidermis. Dev. Biol. 46: 436-438.

Blanck, C. E., McAleese, S. R., and Sawyer, R. H. (1981). Morphogenesis of conjunctival papillae from normal and scaleless chick embryos. Anat. Rec. 199:249-257.

Brotman, H. F. (1977a). Epidermal-dermal tissue interactions between mutant foot skin and normal back skin: A comparison of the inductive capacities of scaleless low line and normal anterior foot dermis. J. Exp. Zool. 200:125-136.

Brotman, H. F. (1977b). Epidermal-dermal tissue interactions between mutant and normal embryonic back skin: Site of mutant gene activity determining abnormal feathering in the epidermis. J. Exp. Zool. 200:243-258.

Chirgwin, J. M., Przbyla, A. E., MacDonald, R. J., and Rutter, W. J. (1979). Isolation of biologically active ribonucleic acid from sources enriched in ribonuclease. Biochemistry 18:5294-5299.

Coulombre, J. L. and Coulombre, A. J. (1971). Mediplastic induction of scales and feathers in the corneal anterior epithelium of the chick embryo. Dev. Biol. 25:464-478.

Dhouailly, D. (1975). Formation of cutaneous appendages in dermo-epidermal recombinations between reptiles, birds and mammals. Wilhelm Roux's Archives 177:323-340.

Dhouailly, D. (1978). Feather-forming capacities of the avian extra-embryonic somatopleure. J. Embryol. Exp. Morphol. 43:279-287.

Dhouailly, D., Rogers, G. E., and Sengel, P. (1978). The specification of feather and scale protein synthesis in epidermal-dermal recombinations. Dev. Biol. 65:58-68.

Engrall, E., Ruoslahti, E., and Miller, E. (1978). Affinity of fibronectin to collagens of different genetic types and to fibrinogen. J. Exp. Med. 147:1584-1595.

Fisher, C. and Sawyer, R. H. (1979). Response of the avian chorionic epithelium to presumptive scale-forming dermis. J. Exp. Zool. 207:505-512.

Goetinck, P. F. (1980). Genetical aspects of skin differentiation. In The Skin of Vertebrates (R. I. C. Spearman and P. A. Riley, eds.), Academic Press, New York.

Goetinck, P. F. and Abbott, U. K. (1963). Tissue interactions in the scaleless mutant and the use of scaleless as an ectodermal marker in studies of normal limb differentiation. J. Exp. Zool. 154:7-19.

Goetinck, P. F. and Sekellick, M. J. (1970). Early morphogenetic events in normal and mutant skin development in the chick embryo and their relationship to alkaline phosphatase activity. Dev. Biol. 21:349-363.

Goetinck, P. F. and Sekellick, M. J. (1972). Observations on collagen synthesis, lattice formation and morphology of scaleless and normal embryonic skin. Dev. Biol. 28:636-648.

Gold, L. and Pearlstein, E. (1980). Fibronectin-collagen binding and requirement during cellular adhesions. Biochem. J. 186: 551-559.

Haake, A. R. and Sawyer, R. H. (1982). Avian feather morphogenesis: Fibronectin-containing anchor filaments. J. Exp. Zool. 221: 119-123.

Kallman, F., Evans, J., and Wessells, N. K. (1967). Anchor filament bundles in embryonic feather germs and skin. J. Cell Biol. 32:236-240.

Kato, Y. (1969). Epithelial metaplasia induced on extraembryonic membranes. I. Induction of epidermis from chick chorionic epithelium. J. Exp. Zool. 170:229-252.

Kato, Y. and Hayashi, Y. (1963). The inductive transformation of the chorionic epithelium into skin derivatives. Exp. Cell Res. 31:599-602.

Kemp, D. J. (1975). Unique and repetitive sequences in multiple genes for feather keratin. Nature 254:573-577.

Kemp, D. J. and Rogers, G. E. (1972). Differentiation of avian keratinocytes: Characterization and relationship of the keratin proteins of adult and embryonic feathers and scales. Biochemistry 11:969-975.

Kischer, C. W. and Keeter, J. S. (1973). Anchor filament bundles in embryonic skin: Origin and termination. Am. J. Anat. 130: 179-194.

Knapp, L. W. and Sawyer, R. H. (1982a). Immunohistochemical localization of alpha-keratin and fibronectin in skin cell aggregates. Trans. Am. Microsc. Soc. (in press).

Knapp, L. W. and Sawyer, R. H. (1982b). Difference between the normal and scaleless (sc/sc) chicken epidermal cell surface detected with wheat germ agglutinin. Exp. Cell. Res. 139:(in press).

Kollar, E. J. (1981). Tooth development and dental patterning. In Morphogenesis and Pattern Formation (T. G. Connelly, L. L. Brinkley, and B. M. Carson, eds.), pp. 87-102. Raven Press, New York.

Lash, J. W. and Burger, M. M. (1977). Cell and Tissue Interactions. Society of General Physiologists Series, Vol. 32, Raven Press, New York.

Linsenmayer, T. F. (1972). Control of integumentary patterns of the chick. Dev. Biol. 27:244-271.

Lockett, T. J., Kemp, D. J., and Rogers, G. E. (1979). Organization of the unique and repetitive sequences in feather keratin messenger ribonucleic acid. Biochemistry 18:5654-5663.

Lucas, A. M. and Stettenheim, P. R. (1972). Avian Anatomy-Integuments. U.S. Government Printing Office, Washington, D.C.

Maderson, P. F. A. (1965). The embryonic development of the squamate integument. Acta Zool. 46:275-295.

Maderson, P. F. A. (1972). Some speculations on the evolution of the vertebrate integument. Am. Zool. 12:159-171.

Maderson, P. F. A. and Sawyer, R. H. (1979). Scale embryogenesis in birds and reptiles. Anat. Rec. 193:609 (Abstract).

McAleese, S. R. and Sawyer, R. H. (1981). Correcting the phenotype of the epidermis from chick embryos homozygous for the gene scaleless (sc/sc). Science 214:1033-1034.

McAleese, S. R. and Sawyer, R. H. (1982). Avian scale development. IX. Scale formation by scaleless (sc/sc) epidermis under the influence of normal scale dermis. Dev. Biol. 89:493-502.

Mirkes, P. E. and Sawyer, R. H. (1979). Abnormal scale morphogenesis: Relationship of polypeptide pattern to action of the scaleless gene. J. Exp. Zool. 208:195-200.

Moscona, A. (1959). Squamous metaplasia and keratinization of chorionic epithelium of the chick embryo in egg and in culture. Dev. Biol. 1:1-23.

Moscona, A. (1960). Metaplastic changes in the chorioallantoic membranes. Transpl. Bull. 26:120-124.

Overton, J. and Collins, J. (1976). Scanning electron microscopic visualization of collagen fibers in embryonic chicken skin. Dev. Biol. 48:80-90.

O'Guin, W. M., Knapp, L., and Sawyer, R. H. (1982). Biochemical and immunohistochemical localization of alpha and beta keratins in avian scutate scale. J. Exp. Zool. 220:371-376.

O'Guin, W. M. and Sawyer, R. H. (1982). Avian scale development. VIII. Relationships between morphogenetic and biosynthetic differentiation. Dev. Biol. 89:485-492.

Palmoski, M. J. and Goetinck, P. F. (1970). An analysis of the development of conjunctival papillae and scleral ossicles in the eye of the scaleless mutant. J. Exp. Zool. 174:157-164.

Parakkal, P. F. and Alexander, N. J. (1972). Keratinization, A Survey of Vertebrate Epithelia. Academic Press, New York.

Rawles, M. E. (1963). Tissue interactions in scale and feather development as studied in dermal-epidermal recombinations. J. Embryol. Exp. Morphol. 2:765-789.

Sawyer, R. H. (1972a). Avian scale development. I. Histogenesis and morphogenesis of the epidermis and dermis during formation of the scale ridge. J. Exp. Zool. 181:365-381.

Sawyer, R. H. (1972b). Avian scale development. II. A study of cell proliferation. J. Exp. Zool. 181:385-408.

Sawyer, R. H. (1975). Avian scale development. V. Ultrastructure of the chorionic epithelium induced by anterior shank dermis from the scaleless mutant. J. Exp. Zool. 191:133-139.

Sawyer, R. H. (1978). Keratogenic metaplasia of the avian chorionic epithelium: Absence of the beta stratum which characterizes the epidermis of the avian scutellate scale. J. Exp. Zool. 205:225-242.

Sawyer, R. H. (1979). Avian scale development: Effect of the scaleless gene on morphogenesis and histogenesis. Dev. Biol. 68:1-15.

Sawyer, R. H. and Abbott, U. K. (1972). Defective histogenesis and morphogenesis in the anterior shank skin of the scaleless mutant. J. Exp. Zool. 181:99-110.

Sawyer, R. H., Abbott, U. K., and Fry, G. N. (1974a). Avian scale development. III. Ultrastructure of the keratinizing cells of the outer and inner epidermal surfaces of the scale ridge. J. Exp. Zool. 190:57-70.

Sawyer, R. H., Abbott, U. K., and Fry, G. N. (1974b). Avian scale development. IV. Ultrastructure of the anterior shank skin of the scaleless mutant. J. Exp. Zool. 190:71-78.

Sawyer, R. H., Abbott, U. K., and Trelford, J. D. (1972). Inductive interactions between human dermis and chick chorionic epithelium. Science 175:527-529.

Sawyer, R. H. and Borg, T. K. (1979). Avian scale development. VI. Ultrastructure of the keratinizing cells of reticulate scales. J. Morphol. 161:111-122.

Sawyer, R. H. and Borg, T. K. (1980). Avian scale development. VII. Normal keratinization follows abnormal morphogenesis of reticulate scales from the scaleless mutant. J. Morphol. 166: 197-202.

Sawyer, R. H. and Craig, K. F. (1977). Avian scale development: Absence of an "epidermal placode" in reticulate scale morphogenesis. J. Morphol. 154:83-94.

Saxen, L. and M. Karkinen-Jääskeläinen (1981). Biology and pathology of embryonic induction. In Morphogenesis and Pattern Formation (T. G. Connelly, L. L. Brinkley, and B. M. Carson, eds.), pp. 21-48, Raven Press, New York.

Sengel, P. (1958). Recherches experimentales sur la differenciation des germes plumaires et du pigment de la peau de l'embryon de poues en culture in vitro. Ann. Sci. Nat. Zool. 20:431-514.

Sengel, P. (1976). Morphogenesis of skin. In Developmental and Cell Biology Series (M. Abercrombie, D. R. Newth, and J. G. Torrey, eds.), Cambridge University Press, Cambridge.

Sengel, P. and Abbott, U. K. (1963). In vitro studies with the scaleless mutant: Interaction during feather and scale differentiation. J. Hered. 54:254-262.

Slavkin, H. C., Trump, G. N., Brownell, A., and Sorgente, N. (1977). Epithelial-mesenchymal interactions: Mesenchymal specificity. In Cell and Tissue Interactions, Society of General Physiologists Series, Vol. 32 (J. W. Lash and M. Burger, eds.), pp. 29-46. Raven Press, New York.

Smoak, K. D. (1980). The avian metatarsal spur: Morphogenesis and keratinization in normal and "scaleless" chicks. Masters Thesis, University of South Carolina, Columbia, S.C.

Smoak, K. D. and Sawyer, R. H. (1982). Avian spur development: Abnormal morphogenesis and keratinization in the scaleless (sc/sc) mutant. Trans. Am. Micros. (in press).

Spearman, R. I. C. (1966). The keratinization of epidermal scales, feathers and hairs. Biol. Rev. 41:59-96.

Spearman, R. I. C. (1973). The Integument, A Textbook of Skin Biology, Cambridge University Press, Cambridge.

Stuart, E. S. and Moscona, A. A. (1967). Embryonic morphogenesis: Role of fibrous lattice in the development of feathers and feather pattern. Science 157:947-948.

Thomas, P. S. (1980). Hybridization of denatured RNA and small DNA fragments transferred to nitrocellulose. Proc. Natl. Acad. Sci. USA 77:5201-5205.

Toole, B. P. and Lowther, D. A. (1968). The effect of chondroitin sulfate-protein on the formation of collagen fibrils in vitro. Biochem. J. 109:857-866.

Toole, B. P., Okayma, O., Orkin, R. W., Yoshimura, M., Muto, M., and Kaji, A. (1977). Developmental roles of hyaluronate and chondroitin sulphate proteoglycans. In Cell and Tissue Interactions (J. W. Lash and M. M. Burger, eds.), pp. 139-154. Raven Press, New York.

Watson, M. (1968). A histological study of feather induction from chick chorionic epithelium. Masters Thesis, University of Massachusetts, Amherst, Massachusetts.

Wessells, N. K. (1965). Morphology and proliferation during early feather development. Dev. Biol. 12:131-153.

Wessells, N. K. (1977). Tissue Interactions and Development, Benjamin/Cummings, Menlo Park, California.

Yamada, K., Schlesinger, D. H., Kennedy, D. W., and Pastan, I. (1977). Characterization of a major fibroblast cell surface glycoprotein. Biochem. 16:5552-5559.

Yohro, T. (1969). The mitotic pattern of the embryonic epidermis of chick during scale morphogenesis. J. Embryol. Exp. Morphol. 21:235-242.

# 7

# FEATHER FORMING PROPERTIES OF THE FOOT INTEGUMENT IN AVIAN EMBRYOS

## D. Dhouailly and P. Sengel

Heterospecific dermal–epidermal recombinations between skin tissues of the three classes of amniotes, reptiles, birds, and mammals have shown that the specific quality of their characteristic cutaneous appendages—namely reptilian scale, feather, and hair—is determined by the epidermis, but that their morphogenesis requires at least two categories of messages originating from the dermis (Sengel, 1976; Dhouailly, 1977a). The early ones, which can be interpreted by an epidermis from another zoological class, are responsible for the initiation of appendage morphogenesis and for their distribution pattern. The later ones, which can only be interpreted by an epidermis from the same zoological class, contain indispensable information for the structural organization, histogenesis, and accomplishment of the reptilian scale, the feather, and the hair. Thus, lizard epidermis forms scale buds, which do not develop to completion and which are arranged like feather buds, mouse pelage hair follicles, or vibrissae, when it is associated respectively with a dorsal chick dermis, and a dorsal or upper-lip mouse dermis. In homospecific recombinations lizard epidermis forms complete scales, large rectangular overlapping ones or small tubercular nonoverlapping ones according to the regional origin of the lizard dermis (Dhouailly, 1975). Likewise, arrested stage 2 or 3 hair follicles are formed in recombinants of mouse epidermis and chick dermis (Dhouailly, 1973), whereas, in the same culture conditions, homospecific mouse heterotopic recombinants differentiate stage 5 or 6 pelage hair follicles or whisker follicles according to the regional origin of the dermis (Dhouailly, 1977b). When dorsal chick epidermis is associated with mouse dermis, the resultant appendages are arrested feathers, the diameter and arrangement of which are in conformity with the regional origin of the dermis (Dhouailly, 1973).

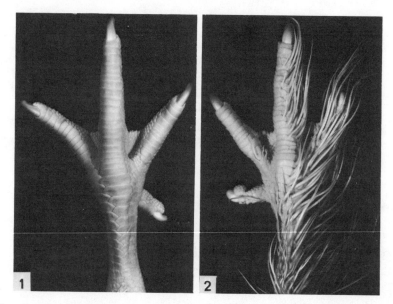

Figures 7.1-7.4 Anterior face of chick feet to compare a normal "clean" foot condition (Fig. 7.1) with three different types of ptilopody (feathered feet). (x 3.33)

Figure 7.1 Large scales or scuta cover the tarsometatarsus and the toes in a 17-day Wyandotte × Rhode Island Red hybrid embryo. Part of the small tubercular reticula which cover the posterior face of the toes and the interdigital membrane can be seen.

Figure 7.2 Genetic ptilopody of a 15-day Bantam of Pekin embryo. Feathered scales are principally located on digit III and distal part of tarsometatarsus, while feathered skin covers digit IV and proximal part of tarsometatarsus.

These arrested feathers possess hypomorphic barb ridges, disposed in an anarchic fashion. Heterospecific dermal-epidermal recombinants between chick and duck embryos reveal that the number and arrangement of barb ridges, and consequently the differentiation of a feather bud into a complete feather, are determined by the dermis (Dhouailly, 1967). When tarsometatarsal chick epidermis is recombined with mouse dermis, rounded buds are formed, the thickness of the epidermis of which remains roughly constant and never form barb ridges. Thus chick epidermis from morphologically undifferentiated skin of the back of a 5-day embryo and that of the shank of a 10-day embryo appear to be nonequivalent. Indeed in birds, the question of the differentiation of cutaneous appendages is complicated by

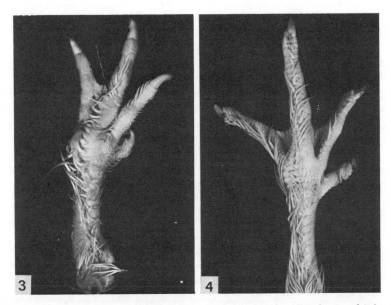

Figure 7.3 Recombinational ptilopody of a 17-day Leghorn chick embryo, which developed from the heterotopic recombination, at 3.5 days of incubation, of wing bud ectoderm, and leg bud mesoderm. Note that scuta on digits II and III and tarsometatarsus bear large feathers on their free edge or are replaced by feathers while reticula are replaced by small and thin feather filaments.

Figure 7.4 Chemical ptilopody in a 17-day chick embryo (Wyandotte × Rhode Island Red hybrid), after a single intraamniotic injection of 125 μg of retinoic acid at 10 days of incubation. Note that the distribution of feather filaments on the scuta is similar to that of genetic ptilopody, and particularly that the largest feathers are borne by digit IV and external row of the tarsometatarsal scuta.

the fact that the epidermal cells can give rise to two types of appendages, namely scales and feathers.

In fowl, most of the body is covered with feathers which are distributed according to a hexagonal pattern, each feather being surrounded by six more or less equidistant feathers. However in common breeds (such as Leghorn, Wyandotte, or Rhode Island) the feet are covered by scaly skin (Fig. 7.1). Large distally overlapping scales (scuta) are arranged in two longitudinal rows on the anterior face of the tarsometatarsus and in a single row on the upper face of the toes. Smaller proximally overlapping scales (scutella) are arranged in two longitudinal rows on the posterior face of the tarsometatarsus. The rest of the foot is covered by small tubercular nonoverlapping rounded scales (reticula).

Feathered feet, so called ptilopody (Danforth, 1919), occurs in relatively few breeds (for example, Brahmas, Silkies, or Bantam of Pekin). Some of the scuta (Fig. 7.2) and scutella may be missing and replaced by feathers or may be maintained and carry a feather at their free edge. A slight manifestation of ptilopody may occasionally (1 percent) occur in breeds of fowl characterized by feet without feathers.

The region-specific determination of cutaneous appendages in birds has been analyzed for many years by heterotopic dermal-epidermal recombinations (Sengel, 1958; Rawles, 1963; Sengel and Pautou, 1969; Kato, 1969; Fisher and Sawyer, 1979). The results show that ectodermal epithelia from any body region or extraembryonic area can be induced to form feathers or scales, depending on the type and stage of dermal mesenchyme they are associated with. Thus, 10- or 11-day anterior tarsometatarsal epidermis will form feathers in combination with 7-day dorsal dermis. Likewise, the association of 3.5-day leg bud ectoderm and wing bud mesoderm leads to the formation of a wing covered with feathers only. The reverse combination of feather-forming epidermis and scale-forming dermis may form typical scales, but only if the tarsometatarsal dermis is obtained from at least 13-day embryos. Indeed, recombinants comprising 9- to 11-day tarsometatarsal dermis and dorsal epidermis develop feathers only and those with 12-day tarsometatarsal dermis both feathers and scales—i.e., feathered scales. Feathered scales or feathers only are also obtained in legs resulting from the combination of 3.5-day leg mesoderm and wing ectoderm (Sengel and Pautou, 1969) (Fig. 7.3). It should be noted that, in all cases, the regional pattern is specified by the dermis, and particularly that the distribution of feathers in recombinants involving foot dermis is a typical scale pattern.

Formation of feathers in heterotopic recombinants of feather-forming epidermis and early scale-forming dermis has until now been interpreted as follows: Foot dermis is endowed with the capacity to induce feathers if it is obtained from embryos before 13 days of incubation. It acquires the capacity to induce scales after 12 days of incubation, due to its contact with foot epidermis. However, there is an apparent contradiction in the fact that tarsometatarsal epidermis is perfectly able to respond to feather-forming instructions originating from feather-forming regions, while it appears unable to respond to the supposed feather-forming message of early tarsometatarsal dermis (Sengel, 1980).

Recent results from chemically-induced ptilopody, from heterotopic recombinants comprising epidermis from a featherless region, and from studies with genetic ptilopody lead to a better understanding of the morphogenetic performance of dorsal-tarsometatarsal recombinants. They offer new arguments for the clarification of the question of regional specification in bird skin.

## RETINOIC ACID-INDUCED PTILOPODY

It has been shown (Dhouailly and Hardy, 1978) that a simple molecule (retinoic acid) when injected (125 μg) at an appropriate stage (10 to 12 days of incubation) into the amniotic cavity of the chick embryo, may bring about changes in the regional specificity of cutaneous appendages of the chick foot in breeds with "clean" feet. These changes (Fig. 7.4) mimic those observed in genetic ptilopody (Fig. 7.2) or in heterotopic limb bud recombinants (Fig. 7.3).

A strict correspondence may be established between the time sequence of appearance of scales (Sawyer, 1972; Sawyer and Craig, 1977) and their sensitivity of the retinoid. In the normal embryo, the large anterior tarsometatarsal scuta appear first, during the 10th day of incubation; posterior scutella form later, on the 11th day; and reticula form last, between the 12th and the 14th day. Correspondingly, only those scuta and scutella which become discernible as opaque patches, or reticula which appear as elevations, during the 24 hours following injection, are affected by the retinoic acid treatment (Dhouailly, Hardy, and Sengel, 1980).

The early effects of retinoic acid on the chick foot integument have been studied (Dhouailly, 1980) after a single injection on day 10, which causes the formation of feathers on the tarsometatarsal scuta. At 11 days (Fig. 7.5), i.e., 24 hours after the injection, in the normal sham-treated embryos, the scale primordia appear as rectangular structures and have reached the hump stage of scale morphogenesis (Sawyer, 1972). Twenty-four hours after retinoic acid injection, scale rudiments appear subdivided into several smaller structures of more or less circular shape (Fig. 7.6). Each of these new abnormally-shaped placodes, arranged according to the scale type pattern, may be interpreted as a feather rudiment, which has the ability to give rise to a feather. These feathers are borne, seven days later, by the distal tip of scales (Fig. 7.4). At the ultrastructural level, changes corresponding to the destruction of the typical scale-type dermal-epidermal junction are observed 24 hours after retinoic acid treatment (Dhouailly, in preparation). The privileged orientation of tarsometatarsal dermal cell processes (Demarchez, Mauger, and Sengel, 1981) at right angle to the proximal-distal axis of the scale is disrupted and replaced by close parallel apposition of broad dermal cell processes against the basement membrane lamina densa of the epidermis, which is a characteristic feature of feather-forming skin. Furthermore, anchor filaments at right angle to the basement membrane, forming piles of epidermal arches are observed. They are the first indication of feather morphogenesis. Figure 7.7 shows feather buds and scale placodes from a feathered feet breed (Bantam of Pekin) which will be discussed later in this chapter under the heading "Genetic Ptilopody."

In order to find out whether the drug acts on both skin tissues or directly on only one of them, heterotypic dermal-epidermal recombinants were prepared in which one of the two tissues was obtained from embryos treated with retinoic acid 24 hours before recombination, the other tissue being obtained from a normal 10-day embryo (Dhouailly, Cadi, and Sengel, in preparation). Explants comprising retinoic acid-treated epidermis and untreated dermis produce feathered scales (Fig. 7.8), whereas the reverse recombinants of normal epidermis and retinoic acid-treated dermis form scales only (Fig. 7.9). These results show that the treatment hits primarily the

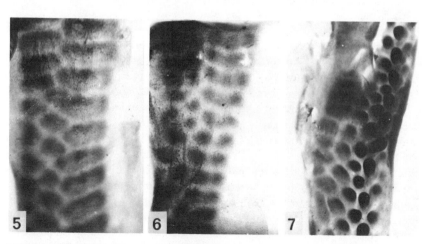

Figures 7.5-7.7 Dissected skin from anterior face of tarsometatarsus of right foot of chick embryos. (x 16)

Figure 7.5 Rectangular scale placodes of an 11-day sham-treated (Wyandotte × Rhode Island Red hybrid) embryo.

Figure 7.6 Circular feather placodes 24 hours after injection at 10 days of retinoic acid to a normal scaled feet breed embryo. Note that these placodes are arranged according to the typical scale pattern.

Figure 7.7 Feather buds and scale placodes of a 10-day embryo from a feathered feet breed (Bantam of Pekin).

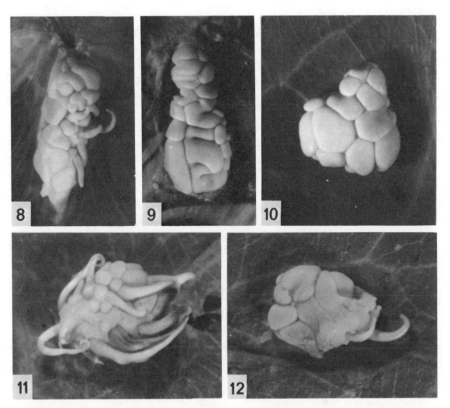

Figures 7.8-7.12 Skin recombinants, cultured for 5 days on the chorioallantoic membrane of the chick embryo. (x 14)

Figure 7.8 Retinoic acid-treated tarsometatarsal epidermis (obtained 24 hours after injection at 10 days) combined with 10-day untreated tarsometatarsal dermis: formation of feathered scales.

Figure 7.9 10-day untreated tarsometatarsal epidermis combined with retinoic acid-treated tarsometatarsal dermis (obtained 24 hours after injection at 10 days): formation of normal scales.

Figure 7.10 10-day midventral apterium epidermis combined with 8.5-day tarsometatarsal dermis: formation of normal scales.

Figure 7.11 8.5-day Bantam of Pekin tarsometatarsal epidermis combined with 8.5-day normal tarsometatarsal dermis: formation of feathered scales (fs) and of feathered skin (f).

Figure 7.12 8.5-day normal tarsometatarsal epidermis combined with 8.5-day Bantam of Pekin tarsometatarsal dermis: formation of scale(s) and of two feathers and five to six arrested feather buds (fb). Note that grafts in Figures 7.11 and 7.12 are shown with the same orientation and result from reciprocal dermal-epidermal recombination of the same two skin samples.

epidermis. Treated epidermis thus becomes temporarily unable to build or maintain scale placodes, which break up into smaller, roundish structures with feather-forming capacity.

As soon as the drug is eliminated from the tissues, which occurs during the third day after treatment (Valquist, Dhouailly, and Pautou, unpublished results), scales resume their development. Furthermore the epidermis-mediated dermal changes are also reversible, provided a normal untreated epidermis is recombined with the treated dermis: in this case scales are formed by 11-day tarsometatarsal dermis, despite its treatment by retinoic acid. This reversion to normality is confirmed by recombinations of treated tarsometatarsal dermis with nontreated epidermis from the featherless 10-day midventral apterium, which does not form cutaneous appendages normally. Recombinants of this kind also form scales.

## HETEROTOPIC RECOMBINANTS

The epidermis of the midventral apterium of a 10-day chick embryo is perfectly able, like 7-day dorsal epidermis, to develop well-formed feathers when it is recombined with 7-day dorsal dermis (Sengel, Dhouailly, and Kieny, 1969).

Heterotopic recombinants of 10-day tarsometatarsal dermis and 10-day midventral epidermis develop normal scales, which start to develop after 2 days of culture on the chick chorioallantoic membrane. Likewise, recombinants of 8.5-day tarsometatarsal dermis and 10-day midventral epidermis form normal scales (Fig. 7.10) which start to develop after 3-4 days of culture (Cadi and Dhouailly, in preparation). These results show that tarsometatarsal dermis is able to acquire and express scale-inducing properties when it is associated with "neutral" epidermis from an apterium, and not necessarily, as thought previously, in contact with tarsometatarsal epidermis only.

## GENETIC PTILOPODY

Goetinck (1967) showed that the mutation Brahma, characterized by ptilopody, affects both epithelial and mesenchymal cells. When ectodermal and mesodermal leg bud components are exchanged between normal and mutant 3.5-day embryos, the resulting legs always display the ptilopody character, whether it is the mesoderm or the ectoderm which originates from the mutant embryo. The same result is obtained with limb recombinants of normal and Bantam of Pekin 3.5-day embryos (Cadi and Dhouailly, in preparation).

In the case of the Bantam of Pekin breed, the morphogenesis of foot skin cutaneous appendages has been followed during normal development. Feather placodes appear first, at 8.5 days of incubation, and scale placodes later, from 10 days onward (Fig. 7.7). Thus, feather placodes are formed in foot skin before the appearance of scales and concomitantly with wing feather placodes.

Dermal-epidermal recombinants between 8.5-day tarsometatarsal tissues of normal and Bantam breeds (Cadi and Dhouailly, in preparation) develop numerous and well-formed feathers in recombinants comprising Bantam epidermis (Fig. 7.11) and no or few feathers (Fig. 7.12) in recombinants comprising Bantam dermis.

## DISCUSSION

Our first interpretation of the retinoic acid effect on the chick foot integument was that this drug could reinforce the presumably weak feather message of the tarsometatarsal dermis or that it could enhance the sensitivity of the tarsometatarsal epidermis to this message (Sengel, 1980). It is clear now that the drug acts primarily on epidermal cells and that these cells are sensitive to the drug only at the time when they are about to start scale placode morphogenesis.

What is the effect of the drug on epidermal cells? It appears, from experiments with epithelial cells from various embryonic and extraembryonic origins, that the responsiveness or competence of the ectodermal epithelium plays an important role in determining whether scales or feathers will be formed. Thus retinoic acid might change the responsiveness of the tarsometatarsal epidermis, by weakening or canceling its ability to form and maintain scale placodes, thus favoring the formation of feather placodes in this bipotential epidermis. The underlying dermis then responds to the placode by forming a dermal feather condensation (Dhouailly, in preparation) instead of a scale condensation. It is known indeed, that in normal feather formation the appearance of the epidermal placode precedes that of the dermal condensation (Sengel and Rusaouën, 1969).

The formation of feathers in recombinants of dorsal epidermis and tarsometatarsal dermis (Sengel, 1958; Rawles, 1963) can also be interpreted as resulting from the region-specific type of competence of the epidermis: dorsal chick epidermis is biased toward feather formation, and responds to the permissive tarsometatarsal patterning message by forming feather placodes, arranged in a scale fashion. The results of heterotopic limb bud recombinants (Sengel and Pautou, 1969) also demonstrate that the ectoderm of a feather-forming region, the wing bud, and that of a scale-forming region, the leg bud, are not equivalent in respect to their morphogenetic

competence. Wing ectoderm is able to express feather-forming poten-
tialities in spite of the leg-type morphogenesis imposed on it by the
leg mesoderm; it is predestined to form feathers. Similarly, in
heterospecific recombinants of mouse hair-forming dermis and chick
epidermis, results are different according to the regional origin of
the chick epidermis: short feathers with malformed barb ridges
differentiate with dorsal epidermis, while atypical buds without barb
ridges form with tarsometatarsal epidermis (Dhouailly, 1973). Heterogenetic leg bud recombinants (Goetinck, 1967) show that in the
ptilopody mutation the leg bud ectoderm of 3.5-day embryos is similar to a wing bud ectoderm. The formation of feathers in recombinants of normal leg ectoderm and ptilopody leg mesoderm suggests
that, in young embryos, early mesenchymal messages could confer
feather properties to the ectodermal cells. With age, the dermal
mesenchyme of the ptilopodous mutant seems to lose this capacity
to confer a feather bias to the overlying epithelium, and to retain
only its scale-inducing property. Mauger (1972) has shown that the
dermal regional specificity is acquired by presumptive dermal cells
long before they form the dermis, that is at 2 days of incubation.

The formation of feathered feet after treatment of 6- or 7-day
nonmutant embryos by X-ray irradiation (Puchkov, 1969) or administration of 5-bromodeoxyuridine (Tanaka, Suyihara and Kato, personal communication) could also be explained by a modification of
the peidermal competence, where scale-forming properties would be
suppressed, by a mechanism which could be different from that of
retinoic action.

The formation of scales in recombinants of 8.5-day tarsometatarsal dermis and epidermis from the midventral apterium reinforces
the hypothesis that normal shank dermis is endowed with patterning
properties of the scale-type only, and that the midventral epidermis
is "neutral" and does not possess a bias toward feather formation,
just like normal tarsometatarsal epidermis.

Finally, the morphogenesis of the avian scale appears to result
from three successive dermal-epidermal interactions: (1) Early 10-
day tarsometatarsal dermis induces the formation of epidermal
placodes and determines their characteristic arrangement in a scale
pattern. The answer of the epidermis to this dermal induction is the
formation of placodes. If the epidermis is "neutral" (without bias
toward feather formation), such as tarsometatarsal or midventral
apterium epidermis, it forms and maintains large rectangular scale
placodes; (2) the scale placode in turn induces the dermis to acquire
further scale properties (Fisher and Sawyer, 1979); and (3) 12-day
tarsometatarsal dermal cells induce the formation of the asymmetric
structure of the scale, comprising an outer thick epidermis consti-
tuted by a large number of keratinizing cell layers and an inner thin

epidermis which forms the junction with the next scale. This second dermal message cannot be interpreted by mammalian epidermis (Dhouailly, 1973), but it can by chick chorionic epithelium (Kato, 1969) provided that step 2 has been accomplished. However, although no placode stage has been recognized in reptiles (Maderson, 1965), perfect scales are formed in recombinants of lizard epidermis and chick tarsometatarsal dermis from 9- to 13-day embryos (Dhouailly, 1975).

It should be noted that the expression of feather properties seems to be easier than that of scale properties by extraembryonic epithelial cells; indeed, these cells are able to acquire and express feather properties by contact with a feather-forming dermis (Kato, 1969; Kieny and Brugal, 1977), but are unable to express scale informations by recombination with early scale-forming dermis (Fisher and Sawyer, 1979). Moreover, feather-like abnormal structures have been observed in recombinants of this last kind (Fisher and Sawyer, 1979) as well as in those of chick chorionic epithelium and mouse upper-lip dermis (Dhouailly, 1978). In these two cases, however, no outgrowing feather buds are formed.

CONCLUSION

In conclusion, the acquisition of regional specificity by the avian integument can be understood as proceeding in two successive binary choices. Early in development, probably at the time of or even before gastrulation, presumptive ectoderm becomes subdivided into two categories; embryonic ectoderm (later epidermis) with full appendage-forming competence, and extraembryonic ectoderm (later amniotic and chorionic epithelium) with a diminished competence, restricted to feather placode formation. In extraembryonic epithelium therefore, the ability to form scale placodes is lost but the capacity to respond to the late histogenetic dermal message (step 3 of scale development) is maintained. Embryonic epidermis becomes then further subdivided into two new categories: (1) epidermis which becomes underlaid by feather-tract dense predermis acquires a predominant competence toward feather formation, and becomes able to respond to any nonspecific but patterned permissive dermal stimulus by forming a feather placode; and (2) epidermis which does not come to lie over a feather-tract predermis remains "neutral," without bias either toward feather or scale. This is the case of presumptive foot epidermis and presumptive epidermis from glabrous regions.

Dermal messages are transmitted in two successive steps (Sengel, 1976; Dhouailly, 1977a). In step 1, the instruction reads: "form placodes in a region-specific pattern (hexagonal or scale-type, according to the regional origin of dermis)." This instruction does not specify whether a feather or a scale has to be built. It simply specifies the spatial organization of future cutaneous appendage. In step 2, the instruction is appendage-specific, either feather or scale, again according to the regional origin of the dermis, and transfers to the epidermis the necessary information for ordered feather or scale histogenesis and keratinization.

In view of this account of cutaneous appendage morphogenesis, retinoic acid, the genetic defect of ptilopody and possibly also X-ray irradiation, probably affect the tarsometatarsal epidermis by depressing its ability to build rectangular scale placodes, which geometrically appear to be more difficult to construct and maintain than the smaller circular feather placodes. Drugs, X-rays, and ptilopody mutation might weaken the stability of the scale placode, which would then break up into the easier to shape feather placodes.

## ACKNOWLEDGEMENTS

This work was supported by financial help from INSERM (Contrat de recherche libre no. 7915142).

## REFERENCES

Cadi, R. and Dhouailly, D. (In preparation) Tarsometatarsal dermal-epidermal interactions in normal and ptilopodous breeds.

Danforth, C. H. (1919). An hereditary complex in the domestic fowl. Genetics 4:587-596.

Demarchez, M., Mauger, A., and Sengel, P. (1981). The dermal-epidermal junction during the development of cutaneous appendages in the chick embryo. Arch. Anat. Microsc. Morphol. Exp. (in press).

Dhouailly, D. (1967). Analyse des facteurs de la différenciation spécifique de la plume néoptile chez le canard et le poulet. J. Embryol. Exp. Morphol. 18:389-400.

Dhouailly, D. (1973). Dermo-epidermal interactions between birds and mammals: differentiation of cutaneous appendages. J. Embryol. Exp. Morphol. 30:587-603.

Dhouailly, D. (1975). Formation of cutaneous appendages in dermo-epidermal recombinations between reptiles, birds and mammals. Roux Arch. Dev. Biol. 177:323-340.

Dhouailly, D. (1977a). Regional specification of cutaneous appendages in mammals. Wilhelm Roux's Archives 181:3-10.

Dhouailly, D. (1977b). Dermo-epidermal interactions during morphogenesis of cutaneous appendages in amniotes. Front. Matr. Biol. 4:86-121.

Dhouailly. D. (1978). Feather-forming capacities of the avian extraembryonic somatopleure. J. Embryol. Exp. Morphol. 43:279-287.

Dhouailly, D. (1980). Action of retinoic acid on chick cutaneous appendages morphogenesis. J. Invest. Dermatol. 74:455 (Abstract).

Dhouailly, D. (In preparation). Sequential study of tarsometatarsal chick skin after retinoic acid treatment.

Dhouailly, D., Cadi, R., and Sengel, P. (In preparation). Effects of retinoic acid on chick tarsometatarsal dermal-epidermal interactions.

Dhouailly, D. and Hardy, M. H. (1978). Retinoic acid causes the development of feathers in the scale-forming integument of the chick embryo. Wilhelm Roux's Archives 185:195-200.

Dhouailly, D., Hardy, M. H., and Sengel, P. (1980). Formation of feathers on chick foot scales: a stage-dependent morphogenetic response to retinoic acid. J. Embryol. Exp. Morphol. 58:63-78.

Fisher, C. and Sawyer, R. H. (1979). Response of the avian chorionic epithelium to presumptive scale-forming dermis. J. Exp. Zool. 207:505-512.

Goetinck, P. F. (1967). Tissue interactions in the development of ptilopody and brachydactyly in the chick embryo. J. Exp. Zool. 165:293-300.

Kato, Y. (1969). Epithelial metaplasia induced on extraembryonic-membrane. I. Induction of epidermis from chick chorionic epithelium. J. Exp. Zool. 170:229-252.

Kieny, M. and Brugal, M. (1977). Morphogenèse du membre chez l'embryon de poulet. Compétence de l'ectoderme embryonnaire et extraembryonnaire. Arch. Anat. Morphol. Exp. 66:235-252.

Maderson, P. F. A. (1965). The embryonic development of the squamate integument. Acta Zool. 46:275-295.

Mauger, A. (1972). Rôle du mésoderme somitique dans le développement du plumage dorsal chez l'embryon de poulet. II. Régionalisation du mésoderme plumigène. J. Embryol. Exp. Morphol. 28:343-366.

Puchkov, V. F. (1969) (En russe). Développement de plumes à partir des écailles des pattes postérieures, après irradiation par rayons X. Arkh. Anat. Gistol. Embriol. 57:105-111.

Rawles, M. E. (1963). Tissue interactions in scale and feather development as studied on dermal-epidermal recombinations. J. Embryol. Exp. Morphol. 11:765-789.

Sawyer, R. H. (1972). Avian scale development. I. Histogenesis and morphogenesis of the epidermis and dermis during formation of the scale ridge. J. Exp. Zool. 181:365-384.

Sawyer, R. H. and Craig, K. F. (1977). Avian scale development. Absence of an "epidermal-placode" in reticulate scale morphogenesis. J. Morphol. 154:83-94.

Sengel, P. (1958). Recherches expérimentales sur la différenciation des germes plumaires et du pigment de la peau l'embryon de poulet en culture in vitro. Ann. Sci. Nat. Zool. 20:431-514.

Sengel, P. (1976). Morphogenesis of skin. In Developmental and Cell Biology Series (M. Abercrombie, D. R. Newth, and J. G. Torrey, eds.), pp. 1-277. Cambridge University Press, Cambridge.

Sengel, P. (1980). Developmental properties of the foot integument in avian embryos. In Teratology of the Limbs. De Gruyter, Berlin.

Sengel, P., Dhouailly, D., and Kieny, M. (1969). Aptitude des constituants cutanés de l'aptérie médio-ventrale à former des plumes. Dev. Biol. 19:436-446.

Sengel, P. and Pautou, M. P. (1969). Experimental conditions in which feather morphogenesis predominates over scale morphogenesis. Nature 222:693-694.

Sengel, P. and Rusaouën, M. (1969). Modifications ultrastructurales au cours de l'histogenèse de la peau chez l'embryon de poulet. Arch. Anat. Microsc. Morphol. Exp. 58:77–96.

# 8
# VITAMIN A AND THE EPITHELIAL-MESENCHYMAL INTERACTIONS IN SKIN DIFFERENTIATION

## M. H. Hardy

DEDICATION

This paper is dedicated to Margaret R. Murray on the occasion of her eightieth birthday. Dr. Murray's pioneering contributions through the organotypic culture of neural and other "difficult" tissues, and especially her stimulation and encouragement of younger researchers over the years, have contributed greatly to our understanding of tissue interactions.

The history of the discovery of vitamin A effects on differentiation is full of surprises. More than a quarter of a century ago Fell and Mellanby (1953) wrote of their astonishment when they found that 7-day embryonic chicken epidermis in organ culture in the presence of high levels of vitamin A developed into a ciliated and mucus-secreting epithelium instead of a keratinizing one. Prior to this, everyone thought that the ectoderm was already "determined" for keratinization. It was known that many adult ciliated and mucus-secreting epithelia underwent keratinization in vivo in the absence of vitamin A, but Fell and Mellanby had shown that an excess of vitamin A could push the embryonic epidermis in the opposite direction to a condition known as mucous metaplasia.

Although many investigators have tried to induce a similar mucous metaplasia in mammalian skin with vitamin A, they have been only partially successful. This question will be considered at the end of this chapter. The main concern here is to discuss the effects of excess vitamin A on hair follicles, and to compare them with the epithelial-mesenchymal interactions which occur in normal development.

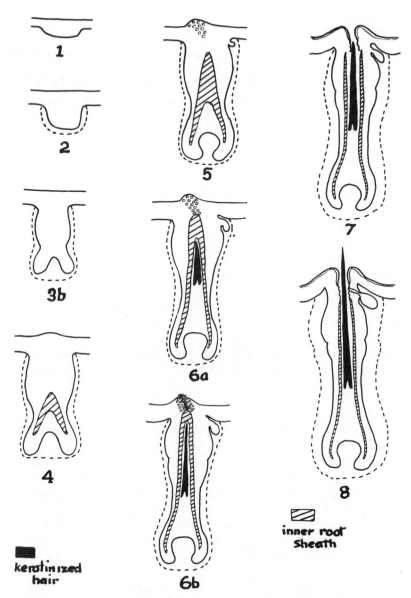

Figure 8.1 Stages in the development of vibrissa follicles in the mouse. In stage 3a, dermal papilla depth < width. In stage 3b, dermal papilla depth = width. In stage 3c, dermal papilla depth > width.

EFFECTS OF VITAMIN A ON HAIR FOLLICLES
IN VITRO

Having read the early papers on hypervitaminosis A and epidermis, I wondered if this condition would affect hair follicles, which
are small organs devoted to the formation of hairs, products which,
unlike the stratum corneum, are almost 100 percent keratins. In
organ cultures of 13.5- to 14-day embryonic mouse skin on a solid
biological medium containing additional retinol (3.8 $\mu$g/ml), I found
that many of the regular hair follicles regressed. However, the large
vibrissa follicles of the snout (Figs. 8.1 and 8.2) gave me my first
surprise, because many of them gave rise to repeatedly branching
and mucus-secreting glands (Fig. 8.3; Hardy, 1968). These glands
arose by budding from the lateral walls of the follicles after about
three days in vitro, but they did not resemble sebaceous glands,
sweat glands, or any skin glands commonly found in mammals. This
time, excess vitamin A had produced not just mucous metaplasia,
but a glandular morphogenesis followed by cytodifferentiation of
mucus-secreting cells. Thus tiny new organs were created, which
were something like immature salivary glands growing in skin. These
experiments have been repeated many times by myself and by gradu-
ate students, and the results were the same with equivalent levels of

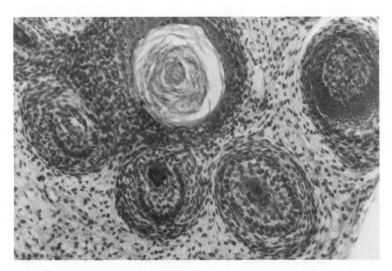

Figure 8.2 Section of explant from upper lip of 14-day mouse embryo after 10 days in standard medium showing central keratinized
epidermis and well developed vibrissa follicles at stage 8, cut at
different depths. Haemalum, eosin, Alcian blue.

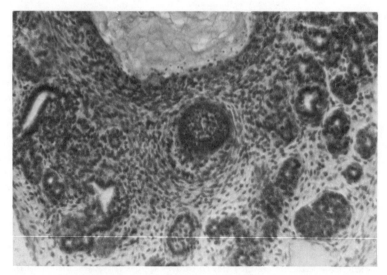

Figure 8.3 Section of explant from upper lip of 14-day mouse embryo after 10 days in vitamin A medium (3.8 μg/ml), showing central, imperfectly keratinized epidermis, one vibrissa follicle at stage 5 and numerous branching glands which have been derived from vibrissa follicles. Haemalum, eosin, Alcian blue.

either retinol or retinyl acetate (Hardy, 1968; Bellows and Hardy, 1977). Explants have been maintained for up to 21 days in vitamin A enriched medium, and the glands continued to grow and differentiate.

A histochemical study with Alcian blue at different $Mg^{2+}$ concentrations, and with the Periodic Acid Schiff method with borohydrite blocking and diastase digestion, confirmed the mucopolysaccharide nature of the secreted material (Bellows and Hardy, 1977). The evidence suggested that the secretory products were either neutral mucosubstances or nonsulfated acidic mucosubstances such as sialomucins.

POSSIBLE MECHANISMS INVOLVED IN
INITIATION OF GLANDS

If excess vitamin A blocked mitosis, inhibited an enzyme involved in follicle differentiation, or interfered with the normal instructive interaction of dermal papillae with follicle hair matrix (the "second message"; Dhouailly, 1975), this might explain the arrest of hair follicle development which has been observed. However, none of these actions could explain the formation of glands de novo. Another mechanism must be sought.

Many teratogens act as cytotoxic agents, affecting a particular type of cell, or cells in general. The result in the former case is usually a malformed structure, such as the open neural tube in hamsters resulting from cell damage in somites and notochord after vitamin A administration to the embryos (Marin-Padilla and Ferm, 1965). If a proportion of cells of all types are damaged, a well-formed but miniature structure may arise from the surviving cells. Neither of these results was obtained when vitamin A additions ranging from 2 to 8 $\mu$g/ml of medium were used in our experiments, nor was there cytological evidence of toxicity, and yet glands were induced by this treatment. A dose-response study (Hardy, 1967) had earlier established that cytotoxic effects did not appear until the dose of retinol reached 11.4 $\mu$g/ml.

Another possibility is that the initiation of glands occurred by a mechanism analogous to a classical epithelial-mesenchymal interaction in which the epithelial cells are induced to embark on a new program of development. Ultrastructural findings, described below, encouraged us to consider this hypothesis.

## ULTRASTRUCTURE OF EPITHELIAL-MESENCHYMAL INTERFACES

It will be necessary to summarize the development of the lateral wall of vibrissa follicles as seen by transmission electron microscopy in vivo, and in standard medium in vitro. The effects of added vitamin A will then be described (Hardy et al., 1982).

The lateral follicle wall of the more advanced vibrissa follicles in 14-day mouse embryos had a continuous basal lamina and a few attachment plaques of future hemidesmosomes. After one day in vitro the basal lamina remained intact, and hemidesmosome plaques were more numerous and had tonofilaments attached (Fig. 8.4). After two days in vitro there was a further increase in the number and complexity of hemidesmosomes beside the intact basal lamina. After three and six days in vitro the basal lamina was still complete, and extracellular amorphous and fibrillar material, including collagen, accumulated lateral to the lamina densa.

When 6 $\mu$g/ml of retinol was added to the standard medium, the findings were very different. After only one day the lamina densa showed ballooning and occasional gaps. Hemidesmosomes were less frequent and less developed than in the standard medium. Mesenchymal cells of the dermal root sheath surrounding the follicle

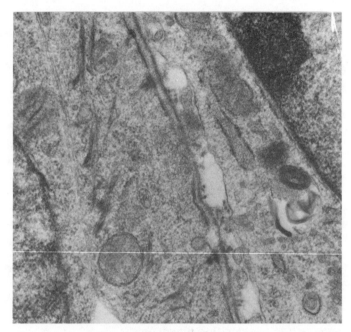

Figure 8.4 Electron micrograph of lateral wall of follicle from 14-day skin after 1 day in standard medium, showing intact basal lamina, and hemidesmosomes with tonofilaments. Some hemidesmosomes show an increased density of the lamina lucida which is due to anchoring filaments.

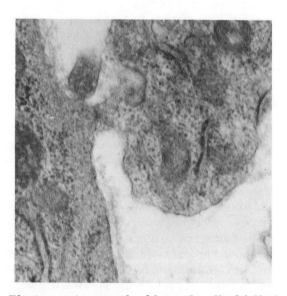

Figure 8.5 Electron micrograph of lateral wall of follicle from 14-day skin after 2 days in medium containing 6 μg/ml retinol, showing a lamina densa of reduced density and variable thickness. Through a small gap in the basal lamina there is contact between the epithelial cell and a mesenchymal cell.

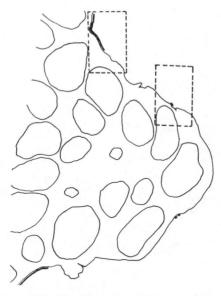

Figure 8.6 Outline of part of a longitudinal section of a vibrissa follicle from a 14-day embryo after 3 days in vitamin A medium (6 μg/ml retinol), showing a large lateral bud. The outline was traced from a montage of 6 electron micrographs. The fine line indicates the plasma membrane of the marginal epithelial cells and the heavy line indicates the presence of a basal lamina. The lamina is continuous only at the margins of the bud. (The approximate boundary of Figure 8.7 is indicated by the box on the left.)

approached the basal lamina region and occasionally made contact with the plasma membrane of an outer root sheath cell. After two days there were frequent contacts of this type through the gaps, which were now common, in the basal lamina (Fig. 8.5). After three days the budding of small groups of cells from the lateral follicle was visible at the light microscopic level (Fig. 8.6). The electron microscope showed that on the growing buds there was little or no basal lamina left, and there were large areas of close contact between the epithelial cells of a bud and the surrounding mesenchyme cells (Figs. 8.6, 8.7). There were no hemidesmosomes on the buds. In the lateral follicle walls between the buds, the basal lamina was continuous and many hemidesmosomes matured.

In 1973, when we first observed these gaps in the basal lamina after one day in vitamin A medium (Hardy et al., 1973) we were very much surprised. It was formerly believed that an intact basal lamina separated epithelium from mesenchyme during all secondary tissue

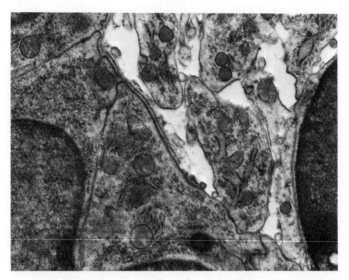

Figure 8.7 Electron micrograph of the margin of the large bud growing from the side of a follicle (Fig. 8.6). In the upper left of the figure is the continuous basal lamina which surrounds the nonbudding part of the lateral follicle wall. In the lower right is the base of the growing bud, where the basal lamina is absent and the epithelial cell makes close contact (12-15 nm) with two portions of dermal cells.

interactions (Hay, 1968). However, a loss of basal lamina and the appearance of direct epithelial-mesenchymal cell contacts at the site of, and at the time of, known epithelial-mesenchymal interactions has since been reported in a number of different organs in vivo or in vitro, such as tooth germs (Slavkin and Bringas, 1976), salivary glands (Cutler, 1977), and lung buds (Bluemink et al., 1976; Wessells, 1977). Furthermore, in an experimental in vitro system, Saxen and Lehtonen (1978) have shown that direct contact between the cells of interacting tissues is essential, at least in the early stages of kidney development.

More recently we have studied the basal lamina region of the dermal papillae in our own cultures of 14-day skin in standard medium. We have found gaps appearing in the basal lamina surrounding the dermal papilla after 2, 3, and 6 days, and contacts through the gaps at 3 and 6 days. These times correspond with the time of the presumed second dermal interaction ("second message") which leads to hair formation. The borders of the dermal papilla and the outer surface of the hair bulb retained an intact basal lamina throughout development. This suggests that the normal interaction of dermal papilla with hair matrix is another one in which heterotypic cell contacts may play a role (Hardy et al., 1982).

It should be noted that it is the <u>sequence</u> of events—(1) breakdown of basal lamina; (2) heterotypic cell contacts; and (3) altered behavior of epithelial cells (whether in mitotic rate, morphogenesis, or cytodifferentiation)—in all the examples quoted above which lends some credence to the idea that they may be causally related.

## DOES THE INITIATION OF MUCOUS GLANDS FROM HAIR FOLLICLES BY VITAMIN A RESULT FROM A TISSUE INTERACTION?

Wessells (1977) has indicated two criteria which can be specified for instructive interactions but can also apply to permissive interactions:

1. In the presence of tissue A, the responding tissue B develops in a specific way.
2. In the absence of tissue A, tissue B fails to develop in that way.

If one substitutes "excess vitamin A" for "tissue A" and "developing vibrissa follicles" for "tissue B," and if one describes the development in a specific way as "undergoes glandular morphogenesis," then both statements are true for the initiation of mucous glands from vibrissa follicles.

Two other characteristics of tissue interactions in early development are:

3. The interactions are possible only at certain stages of development of each tissue (i.e., both tissues must be in the period of "competence").
4. The interaction needs to take place for only a specified period in order to produce an effect.

These two characteristics also apply to the glandular morphogenesis phenomenon, and the evidence to support this will now be presented.

### Relation to Stages of Development

The vibrissa follicles are arranged on the upper lip in one vertical and five horizontal rows (Danforth, 1925). They develop in each horizontal row in a precise caudo-rostral sequence which has been described by Davidson and Hardy (1952), Yamakado and Yohro

(1979), and Van Exan and Hardy (1980). Therefore, in an embryo of any age from 12 days onward, rows of follicles at different stages of development can be seen. In explants of upper lip skin from embryos with gestational ages of 13.0, 13.5, 14.0, and 15.0 days, the follicles destined to show glandular morphogenesis in vitamin A enriched medium were never at less than stage 1, or more than stage 3b (Fig. 8.1; Hardy, 1969) at the time of explantation. Follicles at stage 2 to stage 3a had the highest probability of change. Thus only the more advanced caudal follicles from 13-day embryos and the less advanced rostral follicles from 15-day embryos were affected by vitamin A. It was therefore concluded that the stages from 1 to 3b defined the critical period of competence of vibrissa follicles to respond to vitamin A by glandular morphogenesis. This period occupied about 2.0 to 2.5 days in the life of each follicle.

Relation to Duration of Exposure to Vitamin A

When explants were grown with excess vitamin A for seven days, then washed and transferred to standard medium, the glands which had formed in the first week continued to grow and differentiate during a further week in culture. Explants exposed to the vitamin treatment for only three days behaved similarly when transferred to standard medium for 11 days. In a later series of experiments conducted with Dr. D. Dhouailly, it was found that retinoic acid induced similar changes, but at a more rapid rate than when retinol or retinyl acetate was used. The addition of 5.2 $\mu$g/ml of retinoic acid to explants from 12.5-day embryos for only one day was sufficient to cause glandular morphogenesis in nearly half the explants when they were washed and transferred to a standard medium for five to eight days. Unfortunately we do not yet have data on the concentration of retinoic acid in the explants after their transfer to standard medium, although the compound is known to be unstable, both in vivo and in vitro, unless it is bound to a protein. However, from the above results and from the known properties of retinoids, it seems reasonable to conclude that the high level of one of these compounds, which is required to initiate the glands, is not required to maintain them. Nor was an excess of retinoid necessary for the subsequent cytodifferentiation of mucus-secreting cells. In fact, seven days in vitamin A followed by seven days in standard medium was just as effective as 14 days in vitamin A in producing mucus secretion (Fig. 8.8; Hardy and Bellows, 1978).

Attempts were made to suppress the induction of glands by adding hydrocortisone to the vitamin A medium. However, a dose of 7.5 $\mu$g/ml, which is known to counteract the action of vitamin A on

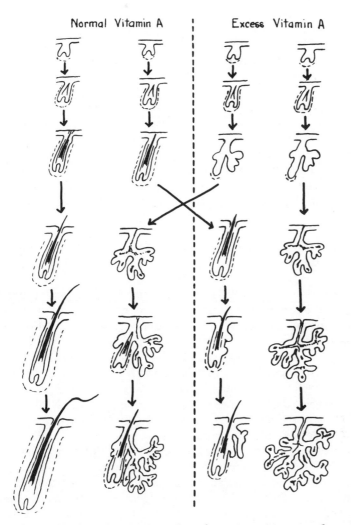

Normal Vitamin A    Excess Vitamin A

Figure 8.8 Diagram showing results of treatment reversal experiment. Column 1 shows the typical appearance of a stage 3b vibrissa
follicle during 14 days in standard medium with a physiologically
"normal" quantity of vitamin A. The upper three sketches in column 2
indicate development during 7 days in standard medium. These explants were transferred to excess vitamin A medium (arrow to column 3), where hair growth was maintained for a further 7 days, but
a few small buds developed, suggesting early gland morphogenesis.
The upper 3 sketches in column 3 indicate glandular morphogenesis
and loss of dermal papillae in excess vitamin A. After transfer to
standard medium (arrow to column 2), some follicles recovered hair
growth but continued glandular morphogenesis and differentiation.
Column 4 shows the effect of 14 days in excess vitamin A.

173

many cultured organs, including chick embryo skin (Fell, 1962; Lasnitzki, 1965), did not suppress glandular morphogenesis or mucous differentiation in the glands (Hardy, 1968). Even a dose of 18 µg/ml of hydrocortisone did not produce a toxic, inhibitory or anti-metaplasia effect on the vibrissa follicles (Hardy and Bellows, 1978).

To sum up, the production of mucous glands by excess vitamin A has four of the features which suggest a tissue interaction. We can also recognize in it a restrictive event, the restriction of lateral follicle cells to branching gland cells, and two expressive events, glandular morphogenesis and mucous metaplasia. So far, both of these expressive events appear to be relatively stable during environmental modifications.

## VITAMIN A AS A MESSAGE ITSELF OR A MESSAGE EVOKER

Whether vitamin A is itself a message to the epithelium, or whether it evokes a message from the dermis has proved to be a very difficult question to answer. Three approaches to it will now be described.

First, we wondered whether the vitamin passed from the medium into the epithelium, and how quickly. The excess vitamin A was added to cock plasma before this was mixed with chicken embryo extract to make the very firm clot of medium, on which the skin explant rested, with the dermis down. Some of the added vitamin A might be bound by retinol binding protein, but most of it would be free in the plasma clot. Since the naturally-occurring, biologically active retinoids are extremely unstable, a marker which would indicate only the intact molecules was sought in preference to a radioactive label on vitamin A. We chose to study the location of the autofluorescence of vitamin A with its excitation and emission characteristics, which correlates well with biological activity. Robert Van Exan found that when 4.1 or 6.9 µg/ml of retinyl acetate was added to the medium, the specific, green, rapidly-fading fluorescence was located predominantly in small lipid droplets, mainly in dermal fibroblasts nearest to the clot, after three days in vitro. After six days in vitro the distribution was similar, except that the fluorescent droplets in fibroblasts were more evenly distributed through the dermis. However, even at three days there was sometimes a faint and diffuse but specific vitamin A fluorescence in the epidermis and the hair follicles. Unfortunately, the distribution of vitamin A fluorescence after one and two days in vitro has not yet been examined, so we do not know whether the breakdown of basal lamina precedes or follows the entry of small quantities of vitamin A into the epithelia. So far, we have failed to show that epithelial tissues are induced before any vitamin A enters their cells, so we cannot assert that modification of the dermis is the only route by which the epithelia are altered.

The second approach was to look for changes in the dermal intercellular matrix in vitamin A medium, since it is known that matrix components alter the growth and differentiation of several epithelia in vitro (e.g., salivary glands; Cohn et al., 1977). Van Exan and I studied the developing dermis in vitro (Van Exan, 1979; Van Exan and Hardy, unpublished observations). During the first few days in the standard medium, the mesenchyme cell density increased around the lateral follicle walls, and collagen fibers were visible in the light microscope. In vitamin A enriched medium the density of mesenchyme cells either remained constant or declined, and collagen fibers were not visible. The transmission electron microscope revealed that in the standard medium fibrils were more frequent, and each fiber was formed by the assembly of fibrils of the same size. In the vitamin A medium there were fewer fibrils, and each fiber was composed of fibrils of different sizes. These observations suggest that vitamin A affects the rates of synthesis and/or breakdown of collagen fibrils and their manner of assembly. The glycosaminoglycans, which are involved with the assembly of fibrils, are also implicated by these findings. In addition, preliminary histochemical observations at the light microscopic level indicated changes in the mucosubstances at the basement membranes of growing glands. Whether any of these changes influence the initiation of glands is not known, but a looser matrix could act permissively to facilitate the extension of budding glands from the follicle walls.

A third way of trying to determine the role of vitamin A was though tissue recombination. The experiments were performed in collaboration with Dr. Danielle Dhouailly in Dr. Philippe Sengel's laboratory at the Université de Grenoble. They are incomplete and still unpublished. We found that retinoic acid was more effective than retinol or retinyl acetate with respect to the speed with which glands were initiated, and for technical reasons it was chosen for these experiments.

Pieces of the dermis from 11.0- to 11.5-day embryonic mice were isolated by trypsin digestion and cultivated for two days, either in the presence or the absence of added retinoic acid. This dermis was then recombined with trypsin-isolated epidermis from 12.5-day untreated mice, and the recombinants were grafted to the chorioallantoic membrane of an 11-day chick for seven or eight days (Fig. 8.9). Of the 20 recovered grafts containing vitamin A treated dermis which developed vibrissa follicles, six also developed a few glands, as illustrated diagrammatically in Figure 8.9. All of the 11 recovered grafts containing untreated dermis developed normal vibrissa follicles, but no glands were found.

Because of difficulties in maintaining the isolated mouse dermis in vitro for two days, the yields of well-differentiated grafts from this type of experiment were very low. Consequently we tested the effect of dermis from 12.2-day chick embryos which had received three daily injections of retinoic acid in vivo. Skin was dissected from the tarsometatarsal region of the chicks, then the dermis was separated from the epidermis with trypsin, thoroughly washed to remove excess retinoic acid, and recombined with untreated 12.5-day mouse epidermis and grafted to the chorioallantoic membrane. Much better yields of differentiated grafts were obtained by this method. These grafts contained hair "buds" (stage 2) but no complete hair follicles because chick dermis does not carry the "second specific dermal message" for hair formation (Dhouailly, 1975). Of the recovered grafts containing retinoic acid treated dermis, 85 percent

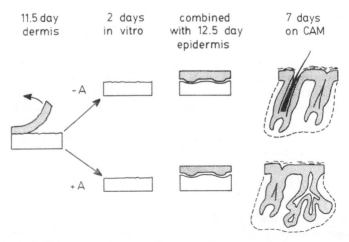

Figure 8.9 Diagram showing the procedure for separating and recombining pieces of dermis and epidermis. The embryonic dermis, separated from epidermis after gentle trypsinization, is grown as organ culture for 2 days, either in standard medium (-A) or with added retinoic acid (+A), then combined with a freshly separated epidermis from an embryo which had not been treated with retinoic acid. The combined tissues are transferred to the chorioallantoic membrane of a chick. Typical results are illustrated. When the dermis was untreated, normal hair follicles only were produced. When the dermis was treated with retinoic acid, some normal hair follicles (early stages only) and some branching glands were formed.

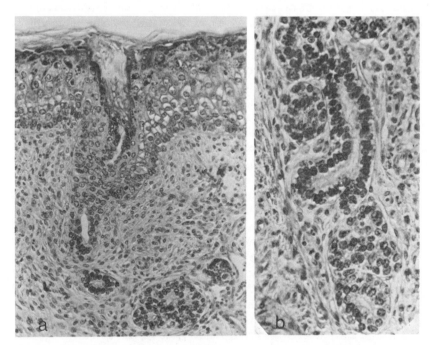

Figure 8.10a Recombinant graft of 12.5-day mouse upper lip epidermis with 12.2-day dermis of a chick which has been injected with retinoic acid. The large (arrested) hair bud has a hair canal. A branching gland has developed from the hair bud.

Figure 8.10b Another section of the gland seen in Figure 10a, showing pale terminal buds and darker duct sections.

(Photographs courtesy of Dr. D. Dhouailly.)

showed well-developed mucous glands arising from the arrested hair buds (Figs. 8.10a and 8.10b).

A conclusion from these results is that dermis of either mouse or chick which has been treated with retinoic acid is capable of supporting glandular morphogenesis of follicle epithelium which has not been treated by a retinoid. The reverse recombinations have not yet been made.

The three groups of studies outlined in this section indicate that vitamin A from the medium is concentrated in the dermis, where it causes rapid changes in collagen and intercellular matrix as well as basal lamina, and that this dermis supports glandular morphogenesis of the untreated follicles or hair buds. Since we have not completely ruled out the possibility that a small quantity of retinoid remaining in the treated dermis might penetrate the epithelium and cause it to initiate glandular morphogenesis, we must for the present refer to the inducing material by the inclusive term "dermis plus vitamin A." (Unfortunately it is not possible to test the effect of vitamin A without any mesenchymal tissue, because neither follicle nor gland morphogenesis can proceed in isolated epidermis.)

## DOES THE VITAMIN A EFFECT MEET THE CRITERIA FOR AN INSTRUCTIVE INTERACTION?

Wessells (1977, p. 46) pointed out that if "in the presence of A, tissue C, normally destined to mature differently, is altered to develop like B, that is, in the specific manner associated with A," the interaction is an instructive one. By this criterion, glands (B) arise from a hair follicle (C) in the presence of dermis plus vitamin A (A), so that dermis plus vitamin A may therefore be regarded as instructive.

Wessells' final criterion for instructive interactions (Wessells, 1977) is that the responding tissue should not develop in the specific way in response to nonspecific stimuli. For glandular morphogenesis, the one exception to this rule that has been found in the author's ten year of experimentation with different media, hormone additives, etc. was that a few chick-mouse recombinants developed glands without the prior injection of retinoic acid into the chicks. Perhaps this was due to the high level of retinol which is normally present in chick embryo tarsometatarsal skin. This level (150 ng/g) was actually higher than the level of retinoic acid (100 ng/g) remaining in the tarsometatarsal skin of the intact chick 24 hours after our standard injection of 125 $\mu$g (Valquist, personal communication). Pending further experimentation, our conclusion that the interaction is instructive must remain a tentative one.

Some further support for an instructive role for "dermis plus vitamin A" comes from experiments with organ cultures where treatments were reversed. It has been pointed out that glands initiated in vitro continued to develop when transferred to standard medium. However, at the same time, the dermal papilla cells of stage 3b follicles, which have already "instructed" the adjacent epithelial cells by the "second dermal message" but have dispersed in the presence of excess vitamin A, can reassemble and once more direct the production of hairs in a standard medium, as shown in Figure 8.8. Had the epithelial hair matrix cells retained their instructions? Continued glandular morphogenesis and recovered follicle morphogenesis frequently took place at different locations in the same follicle.

It is difficult to see how the final dermal environment, which must have been normal enough to allow a blocked instruction of hair matrix by dermal papilla cells to be reestablished, could be without any effect on the abnormal gland development, unless the latter, too, resulted from an instructive interaction rather than a merely permissive one.

## IS THE VITAMIN A EFFECT ON EPIDERMIS IN VITRO ALSO DUE TO AN INSTRUCTIVE INTERACTION?

The explants from mouse embryos which were grown in vitro to study vibrissa follicles also provided material for our studies of epidermis. Excess vitamin A did not stimulate a new pattern of morphogenesis from this layer, but several degrees of metaplasia were reported, depending on the age at which the vitamin was applied.

When 12-day embryo skin was explanted with vitamin A there was a full mucous metaplasia, comparable to that reported in the chick, with a stratified, ciliated, and mucus-secreting epithelium and no signs of keratinization (Figs. 8.11, 8.12, and 8.13; Sweeny and Hardy, 1976). Thirteen-day skin explanted with vitamin A developed some normally keratinizing epithelium, some parakeratotic epithelium with retained nuclei in the stratum corneum, numerous "Alcian-blue-positive bodies" in the stratum granulosum, and some stratified cuboidal epithelium with shedding cells (Hardy, 1967). Histochemical studies indicated that the Alcian-blue-positive bodies contained moderately sulphated acidic glycosaminoglycans, and that the cuboidal epithelium was secreting Alcian-blue-negative, Periodic-Acid-Schiff-positive, neutral mucosubstances or sialomucins (Bellows and Hardy, 1977). Fourteen-day skin with vitamin A showed some parakeratosis, and some shedding cells and many Alcian-blue-positive bodies, but more areas of normal keratinization than were seen in 13-day skin (Hardy, 1967). Fifteen-day skin showed merely some Alcian-blue-positive bodies in relatively normal skin. These results are summarized in Figure 8.14.

A sequential study of ultrastructural changes in the skin explanted at 14 days showed that the addition of vitamin A resulted in gaps in the basal lamina after one day, contacts of dermal cell processes with basal epidermal cells through the gaps after two days, and the appearance of Alcian-blue-positive bodies after three days (Hardy et al., 1978). Changes were taking place in the extracellular matrix of the adjacent dermis which were similar to those summarized above for the region of the vibrissa follicle walls (Van Exan, 1979).

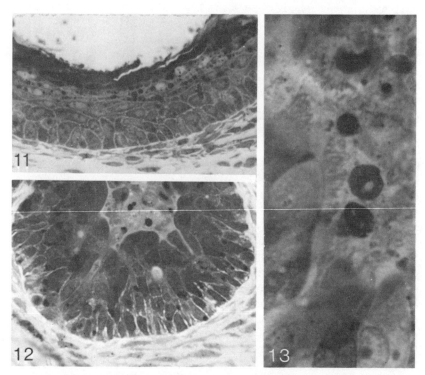

Figure 8.11 One micron Epon section of skin from 12-day embryo after 10 days in standard medium, showing stratified, keratinized epidermis differentiated in vitro. Note the dark superficial flattened cells of the stratum corneum.

Figure 8.12 One micron Epon section of skin from 12-day embryo after 10 days in medium supplemented with 6 μg/ml retinol. The epidermal cells did not become squamous, but retained a polyhedral form. The superficial cells had a rounded border next to an area filled with secreted material and shed cells.

Figure 8.13 High magnification Nomarski interference photomicrograph of a groove in the surface of the epidermis seen in Figure 8.12, showing on the left a ciliated border.

One may argue that Wessells' two criteria for tissue interactions in general were met by adding vitamin A to the medium—in the presence of excess vitamin A, the epidermis produced mucosubstances, and in its absence it did not. The ultrastructural observations provided circumstantial evidence in support of an interaction

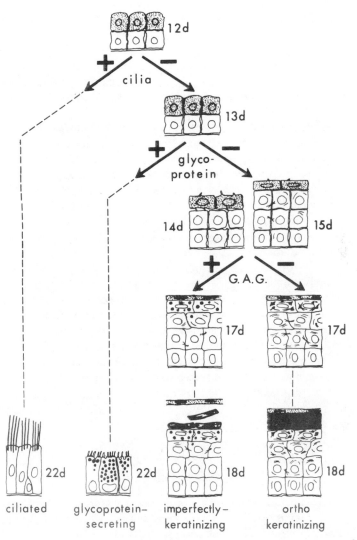

Figure 8.14 Schematic representation showing the fate of embryonic
ectoderm of the mouse treated with excess vitamin A at different
ages (+: treated, -: untreated). Ectoderm treated at 12 days can
develop into ciliated epithelium (and/or glycoprotein secreting). By
13 days cilia are no longer formed, but glycoprotein synthesis is
still possible. By 14 days the stratified, squamous differentiation
has already been determined, but glycoaminoglycan (GAG) synthesis
and imperfect keratinization are possible.

between dermis and epidermis, but in the case of the epidermis, unlike that of the hair follicles, it was not possible to establish a point-to-point relationship between cell contacts and morphological change. The changes were certainly related to stage of development, as were those in vibrissa follicles, but in the epidermis we saw, rather than an all-or-none phenomenon of gland development, a stepwise restriction of potential for cytodifferentiation from day 12 to day 15 (Fig. 8.14).

A fourth property, common to many tissue interactions, which was noted in examining the effect of vitamin A on hair follicles, was irreversibility. This did not seem to apply to the effect on epidermis. A limited time of exposure to the vitamin, such as three days or seven days, was followed by a return of the epidermis to the keratinizing mode when it was transferred to standard medium. In the graft recombinant studies, the 11.5-day mouse epidermis was combined with mouse dermis which had been grown with excess vitamin A. After seven or eight days of growth on the chorioallantoic membrane, there was little or no evidence of retardation of the progress of normal keratinization. The same was true when mouse dermis was combined with chick dermis following injection of the chicks with retinoic acid. These observations support the conclusion of Fell (1957) and others (e.g., McLoughlin, 1961) from their experiments with chicken skin, that the effect of vitamin A on epidermis is a modulating one, but they do not negate the possibility that it is instructive.

Turning now to Wessells' two criteria for instructive interactions, it may be argued: (1) that mucous metaplasia, partial or complete (B), develops from early embryonic epidermis (C) in the presence of dermis plus vitamin A (A), and therefore dermis plus vitamin A is instructive; and (2) that "nonspecific stimuli," such as dermis, or dermis plus hormone, do not produce this result, and therefore dermis plus vitamin A is instructive.

We may tentatively conclude that we are dealing with an instructive interaction acting on an epithelium which is undergoing a progressive series of restrictions in potential. However, the various phenotypes expressed have not been stabilized.


## DOES VITAMIN A INDUCE GLANDULAR MORPHO-GENESIS IN OTHER ECTODERMAL EPITHELIA?

Lawrence and Bern (1960) first reported gland-like downgrowths from adult hamster cheek pouch epithelium when pellets containing high concentrations of vitamin A were implanted in the thin, lightly keratinizing pouches. More recently, Mock and Main (1976) showed that, in organ cultures of newborn hamster cheek pouch, excess vitamin A in the medium induced the formation of very well developed branching and mucus-secreting glands, similar to those reported from mouse vibrissa follicles. Glands or append-ages are not normally associated with the hamster cheek pouch.

In the course of our mouse-mouse and mouse-chick recombi-nation experiments, it was occasionally found that a gland-like struc-ture extended into the grafted dermis from the grafted epidermis at its junction with the chorionic ectoderm.

These two examples suggest that glandular morphogenesis in response to excess vitamin A is not confined to the vibrissa follicles, but may be a property of other ectodermal cells, including the basal cells of some keratinizing epithelia. It would be very interesting to see whether basal lamina gaps and heterotypic cell contacts appeared in the areas of these down growths.

## SIGNIFICANCE OF VITAMIN A INDUCED FEATHER MORPHOGENESIS

It was in the course of preparing retinoid-treated dermis by injecting retinoic acid into chick embryos that Dr. Dhouailly and I discovered that this treatment caused many chickens to grow feathers on the scale-bearing areas of the feet (Dhouailly and Hardy, 1978). (By this time we were used to surprises.) At first this looked as though it could be another instructive interaction of dermis plus vitamin A with foot epidermis. However, the analysis of the results of single injections at different times led to the conclusion that in this case the retinoic acid probably acted at the early scale-forming period by weakening the "make scales and only scales" message which is normally superimposed on the "ubiquitous feather message" in that region. This would therefore not qualify as an instructive interaction, but as a weakening of an interaction, thereby permitting morphologically "normal" structures to appear in an abnormal loca-tion (Dhouailly et al., 1980).

## CONCLUDING REMARKS

There is intense interest in the actions of retinoids at present because they affect development and differentiation, not only in "normal" biological systems, but also in epithelial tumors and dis-eased skin. The mechanisms, however, remain elusive. Retinoids are normally carried in the body to target tissues as retinol, tightly bound to a specific retinol-binding protein. Retinoids have been shown to enter and pass through cell membranes, to bind specific cell surface receptors and cytoplasmic receptors, to cause lysosomal discharge, to enhance glycoprotein synthesis, to alter glycolipids, to stimulate DNA synthesis, and so on. They affect so many cell functions that it is difficult even to select an hypothesis for the mode of action in any tissue, let alone to prove it.

In our experiments it is clear that protein synthesis in the epidermis is being redirected by retinoids. There is strong biochemical evidence that these compounds can stimulate glycosaminoglycan and glycoprotein synthesis in some tissues (De Luca et al., 1979). Hassell and Newsome (1981) have shown that corneal epithelial cells, of the same lineage as epidermal cells, can synthesize "goblet-cell-like" glycoproteins (mucins) under the influence of vitamin A. More recently Fuchs and Green (1981) have produced clear-cut evidence that in their pure human keratinocyte cell cultures, traces of retinyl acetate can regulate the type of keratins synthesized. The 67 kd keratins, which are characteristic of terminally differentiating epidermis, were suppressed by small traces of retinyl acetate in the medium, while 40 kd and 52 kd keratin production was stimulated. Regulation occurred at the level of synthesis, a fact confirmed by translation of keratin-specific mRNA . Interestingly, desquamation and increased cell motility, two features of epidermis exposed to excess vitamin A in other culture systems, were reported by Fuchs and Green to appear only in the presence of retinyl acetate. These experiments suggest that in our more complex cultures, the retinoids, although derived from the dermis, might be operating directly on the epidermal keratinocytes, changing the keratin, glycosaminoglycan, and glycoprotein synthesis in them by biochemical pathways which are now partially characterized.

Less can be said about possible mechanisms for the glandular morphogenesis of vibrissa follicles. Since this process can be initiated by chick dermis plus vitamin A, the mechanism is not class-specific. The process seems analogous to the normal physiological action of steroid hormones on the development of accessory sex organs such as the mammary glands (Kratochwil and Schwartz, 1976), the prostate gland (Cunha, this volume) and the tubular glands of the chick oviduct (Schimke et al., 1975). In some of these systems the steroid acts on the mesenchyme to modify its interaction with the epithelium, and an analogous action of retinoids on dermis seems to be a viable hypothesis for the mechanism of action of retinoids on vibrissa follicles.

While so little is known about the molecules or other signals involved in any tissue interactions, vitamin A has value as a probe for interactions and should continue to play its part by surprising us, challenging our assumptions, and leading us to a better understanding of the normal and the abnormal, both in developing and in mature tissues.

## ACKNOWLEDGEMENTS

The continuing support of the Natural Sciences and Engineering Research Council of Canada is gratefully acknowledged. Some of the research was supported in part by the Ontario Ministry of Agriculture and Food. Much of this work has been a collaborative effort with the coauthors whose papers are referred to in the text, and I thank them for many stimulating discussions. I am indebted to Ms. Carol Ann Thomson for expert typing.

## REFERENCES

Bellows, C. and Hardy, M. H. (1977). Histochemical evidence of mucosubstances in the metaplastic epidermis and hair follicles produced in vitro in the presence of excess vitamin A. Anat. Rec. 187:257-272.

Bluemink, J. G., Van Maurik, P., and Lawson, K. A. (1976). Intimate cell contacts at the epithelial/mesenchymal interface in embryonic mouse lung. J. Ultrastruct. Res. 55:257-270.

Cohn, R. H., Banerjee, S. D., and Bernfield, M. R. (1977). Basal lamina of embryonic salivary epithelia: Nature of the glycosamino-glycans and organization of extracellular materials. J. Cell Biol. 73:464-478.

Cutler, L. S. (1977). Intercellular contacts at the epithelial-mesen-chymal interface of the developing rat submandibular gland in vitro. J. Embryol. Exp. Morphol. 39:71-77.

Danforth, C. H. (1925). Hair in its relation to questions of homology and phylogeny. Am. J. Anat. 36:47-68.

Davidson, P. and Hardy, M. H. (1952). The development of mouse vibrissae in vivo and in vitro. J. Anat. (London) 86:342-356.

De Luca, L. M., Bhat, P. V., Sasak, W., and Adams, S. (1979). Vitamin A and glycoprotein and membrane metabolism. Fed. Proc. 38:2535-2539.

Dhouailly, D. (1975). Formation of cutaneous appendages in dermo-epidermal recombinations between reptiles, birds and mammals. Wilhelm Roux's Arch. 177:323-340.

Dhouailly, D. and Hardy, M. H. (1978). Retinoic acid causes the development of feathers in the scale-forming integument of the chick embryo. Wilhelm Roux's Arch. 185:195-200.

Dhouailly, D., Hardy, M. H., and Sengel, P. (1980). Formation of feathers on chick foot scales: a stage dependent morphogenetic response to retinoic acid. J. Embryol. Exp. Morphol. 58:63-78.

Fell, H. B. (1957). The effect of excess vitamin A on cultures of embryonic chicken skin explanted at different stages of differentiation. Proc. R. Soc. London (Biol.) 146:242-256.

Fell, H. B. (1962). The influence of hydrocortisone on the metaplastic action of vitamin A on the epidermis of embryonic chicken skin in organ culture. J. Embryol. Exp. Morphol. 10:389-409.

Fell, H. B. and Mellanby, E. (1953). Metaplasia produced in cultures of chick ectoderm by high vitamin A. J. Physiol. (London) 119:470-488.

Fuchs, E. and Green, H. (1981). Regulation of terminal differentiation of cultured human keratinocytes by vitamin A. Cell 25: 617-625.

Hardy, M. H. (1967). Responses in embryonic mouse skin to excess vitamin A in organotypic cultures from the trunk, upper lip and lower jaw. Exp. Cell Res. 46:367-384.

Hardy, M. H. (1968). Glandular metaplasia of hair follicles and other responses to vitamin A excess in cultures of rodent skin. J. Embryol. Exp. Morphol. 19:157-180.

Hardy, M. H. (1969). The differentiation of hair follicles and hairs in organ culture. In Advances in Biology of Skin (W. Montagna, ed.), Vol. 9, pp. 35-60. Pergamon Press, Oxford.

Hardy, M. H., and Bellows, C. G. (1978). The stability of vitamin A-induced metaplasia of mouse vibrissa follicles in vitro. J. Invest. Dermatol. 71:236-241.

Hardy, M. H., Sonstegard, K. S., and Sweeny, P. R. (1973). Light and electron microscopic studies of the reprogramming of epidermis and vibrissa follicles by excess vitamin A in organ culture. In Vitro 8:405 (Abstract).

Hardy, M. H., Sweeny, P. R., and Bellows, C. G. (1978). The effects of vitamin A on the epidermis of the fetal mouse in organ culture—an ultrastructural study. J. Ultrastruct Res. 64:246-260.

Hardy, M. H., Van Exan, R. J., Sonstegard, K. S., and Sweeny, P. R. (1982). Basal lamina changes during tissue interactions in hair follicles—an in vitro study of normal dermal papillae and vitamin A-induced glandular morphogenesis. J. Invest. Dermatol. (in press).

Hassell, J. R. and Newsome, D. A. (1981). Vitamin A-induced alterations in corneal and conjunctival epithelial glycoprotein biosynthesis. Ann. NY Acad. Sci. 359:358-365.

Hay, E. D. (1968). Organization and fine structure of epithelium and mesenchyme in the developing chick embryo. In Epithelial-Mesenchymal Interactions (R. Fleischmajer and R. E. Billingham, eds.), pp. 31-55. Williams and Wilkins, Baltimore.

Kratochwil, K. and Schwartz, P. (1976). Tissue interaction in androgen response of embryonic mammary rudiment of mouse: Identification of target tissue for testosterone. Proc. Nat. Acad. Sci. USA 73:4041-4044.

Lasnitzki, I. (1965). The actions of hormones on cell and organ cultures. In Cells and Tissues in Culture, Vol. 1 (E. N. Willmer, ed.), pp. 591-658. Academic Press, New York.

Lawrence, D. J. and Bern, H. A. (1960). Mucous metaplasia and mucous gland formation in keratinized adult epithelium in situ treated with vitamin A. Exp. Cell Res. 21:443-446.

Marin-Padilla, M. and Ferm, V. H. (1965). Somite necrosis and developmental malformations induced by vitamin A in the golden hamster. J. Embryol. Exp. Morphol. 13:1-8.

McLoughlin, C. B. (1961). The importance of mesenchymal factors in the differentiation of chick epidermis. I. The differentiation in culture of the isolated epidermis of the embryonic chick and its response to excess vitamin A. J. Embryol. Exp. Morphol. 9:370-384.

Mock, D. and Main, J. H. P. (1976). The effect of vitamin A on hamster cheek pouch mucosa in organ culture. J. Dent. Res. 58:635-637.

Saxen, L. and Lehtonen, E. (1978). Transfilter induction of kidney tubules as a function of the extent and duration of intercellular contacts. J. Embryol. Exp. Morphol. 47:97–109.

Schimke, R. T., McKnight, G. S., Shapiro, D. J., Sullivan, D., and Palacios, R. (1975). Hormonal regulation of ovalbumin synthesis in the chick oviduct. In Recent Progress in Hormone Research (R. O. Greep, ed.), Vol. 31, pp. 175–209. Academic Press, New York.

Slavkin, H. C. and Bringas, P. (1976). Epithelial-mesenchyme interactions during odontogenesis. IV. Morphological evidence for direct heterotypic cell–cell contacts. Dev. Biol. 50:428–442.

Sweeny, P. R. and Hardy, M. H. (1976). Ciliated and secretory epidermis produced from embryonic mammalian skin in organ culture by vitamin A. Anat. Rec. 185:93–100.

Van Exan, R. J. (1979). An in vitro study of the effects of excess vitamin A on the differentiation of the mammalian dermis. Ph.D. thesis, University of Guelph, Guelph, Ontario.

Van Exan, R. J. and Hardy, M. H. (1980). A spatial relationship between innervation and the early differentiation of vibrissa follicles in the embryonic mouse. J. Anat. (London) 131:643–656.

Wessells, N. K. (1977). Tissue Interactions and Development. W. A. Benjamin, Menlo Park.

Yamakado, M. and Yohro, T. (1979). Subdivision of mouse vibrissae on an embryological basis with descriptions of variations in the number and arrangement of sinus hairs and cortical barrels in BALB/c (nu/+; nude, nu/nu) and hairless (hr/hr) strains. Am. J. Anat. 155:153–174.

# 9

# EPITHELIAL-MESENCHYMAL INTERACTIONS
# IN CARTILAGE AND BONE DEVELOPMENT

## Brian K. Hall

Over the past few years it has become evident that many of
the cartilages and bones in embryonic and adult vertebrates only
begin to differentiate after chondrogenic or osteogenic mesenchymal
or ectomesenchymal* cells have undergone an inductive interaction
with an embryonic epithelium. A list of such cartilages would include:
(1) Meckel's cartilage of the lower jaw; (2) scleral cartilage of the
eye; (3) cartilage of the otic capsule columella and external ear;
(4) the nasal capsular cartilage; (5) vertebral cartilage; (6) limb
cartilages; and (7) various ectopic cartilages and cartilage within
tumors and teratocarcinomas. A list of bones induced by epithelial-
mesenchymal interactions would include: (1) membrane bones of
the lower jaw (the single dentary of mammals and multiple membrane
bones of submammalian vertebrates); (2) the maxilla of the upper jaw;
(3) bones of the roof, floor, and base of the skull; (4) bones of the
secondary palate; and (5) ectopic bones. Clearly, these lists include
virtually all elements of the embryonic facial, cranial, appendicular,
and axial skeletons, the only major exclusions being: (1) subperios-
teal bone which surrounds shafts of long bones; (2) endochondral
bone of the axial and appendicular skeletons; and (3) secondary

---

*All facial and many cranial cartilages and bones arise from
cells derived from the embryonic neural crest (Hall, 1980a). Such
cells are mesenchymal in appearance but are properly called ectomesenchymal to designate their ectodermal origin (Horstadius, 1950).
For simplicity, I shall refer to them as mesenchymal or mesenchyme
throughout this chapter.

cartilages on membrane bones. Although none of these arise because of epithelial-mesenchymal interactions, all arise because of interactions between mesenchyme and an adjacent tissue, either hypertrophic cartilage in (1) and (2), or another skeletal, muscular or connective tissue at an articulation or insertion in the case of (3). Hall (1978a; 1981a; 1981b) and Scott-Savage and Hall (1979; 1980) have reviewed these interactions.

Several aspects of skeletogenic epithelial-mesenchymal interactions have been reviewed over the past four years. Hall (1978a) saw them as providing the basic mechanisms locating skeletal elements within the embryo. Both neural crest and sclerotomal cells undergo extensive cellular migrations before they begin to differentiate into craniofacial and vertebral cartilages and bones. Interactions with epithelia—before, during, or after migration—are crucial for both timing and positioning of these tissues. Vertebral chondrogenesis has been reviewed recently (Hall, 1977). Morriss and Thorogood (1978) and Hall (1980a) have discussed neural crest-derived craniofacial cartilages and bones in some detail. My article provides an overview of all vertebrates while Morriss and Thorogood's discusses mammalian work, especially the use of epithelial-mesenchymal tissue recombinations to identify the site of action of teratogens. Their chapter, plus chapters in the symposium volumes edited by Bergsma (1975), Melnick and Jorgenson (1979), Gorlin (1980), and Pratt and Christiansen (1980) provides an entry into the literature on defective epithelial-mesenchymal interactions as a basis for congenital or teratogen-induced anomalies.

The relative contributions of intrinsic intracellular and extracellular (tissue interaction and extracellular matrix-mediated) control over both differentiation and maintenance of cartilage and bone have also recently been discussed (Hall, 1978a; 1981a). One of the major obstacles in interpreting experiments on epithelial-mesenchymal interactions is our lack of knowledge of the state of determination of mesenchymal cells at the outset of such interaction. However, it is becoming clear that these skeletogenic tissue interactions are permissive (Hall, 1981c; 1981d), i.e., that mesenchymal cells can respond to a variety of epithelia to form cartilages and bones whose histogenesis and morphogenesis is specific to that mesenchyme. Epithelia elicit an already established capability.

Two further reviews have recently been completed. One (Hall, 1981b) discusses approaches, advantages, disadvantages, and general state of the art of cell-tissue interactions as a way to study skeletal development. Epithelial-mesenchymal interactions in normal and mutant embryos, interactions between skeletogenic mesenchyme, nerves, and blood vessels and ectopic skeletogenesis are discussed. The other chapter (Hall, 1982) reviews the available data on tissue interactions and chondrogenesis, covering initiation of chondrogenesis, cartilage as an inducer and the tissue interactions between synovium and cartilage which lead to degradation of the latter.

Clearly, there is bound to be some overlap between the present chapter and these reviews. Consequently, I will not attempt to survey all cartilages and bones in all species in this chapter. Rather, I will attempt to summarize the current state of our knowledge on how epithelial-mesenchymal interactions in cartilage and bone development work. As modes of action vary from site to site, I will concentrate on the few examples where information on the mechanism of action exist. One would like to be able to generalize these mechanisms to all interactions at all sites in all species. However, I emphasize that, despite the temptation, there is no basis for any such generalizations at this time. Our knowledge is just too fragmentary and incomplete.

## CARTILAGE OF LIMB BUDS

The limb bud of the embryonic chick has provided a model for cytodifferentiation of cartilage, muscle, and connective tissue; for pattern formation, morphogenesis, growth, nerve-muscle interaction, programmed cell death, and the action of mutant genes. Some of the vast literature has been discussed by Abbott (1975) and Hall (1978a) and considerably more by Hinchliffe and Johnson (1980).

Limb buds consist of a core of mesenchyme enveloped by an epithelium. In many vertebrates the apical edge of the epithelium thickens into an apical ectodermal ridge (AER), the presence of which is crucial for laying down the limb skeleton in a proximo-distal sequence. The AER of the embryonic chick wing bud first appears at Hamburger-Hamilton (H.H.) stage 17 with H.H. stage 18 (3 days of incubation) being the earliest stage at which limb mesenchyme will form cartilage when isolated in vitro (Searls, 1968; Summerbell, 1974). After H.H. stage 17, the AER controls the pattern of cartilage development. In an elegant series of studies Gumpel-Pinot (1972; 1973; 1980; 1981) has shown that before H.H. stage 17 (i.e., before first outgrowth of the limb bud) epithelium from the limb territory must interact with presumptive limb mesenchyme if cartilage is to differentiate at all. Seventeen percent of H.H. stage 17 limb mesenchyme fails to chondrify unless cultured with epithelium (Table 9.1). This epithelial-mesenchymal interaction acts across both vitelline membrane barriers and across Nucleopore filters of 0.4 or 0.6 $\mu$ porosity and is not site specific—back skin epithelium can substitute for epithelium of the limb bud (Table 9.1). Whether this inductive interaction is mediated by contact of mesenchymal cells with the basement membrane (as postulated by Gumpel-Pinot, 1980) or by a diffusible morphogen remains to be determined.

TABLE 9.1

Epithelial-Mesenchymal Interactions and Chondrogenesis
of the Avian Limb Bud[a]

| Tissue Combination | H.H. Stage of Tissue Isolation | | | |
|---|---|---|---|---|
| | 12-16 | 17 | 18 | 19 |
| Intact limb bud | 77 | 83 | 83 | 97 |
| Isolated mesenchyme | 0 | 17 | 47 | 79 |
| Mesenchyme recombined with epithelium | 82 | 84 | 82 | 100 |
| Mesenchyme recombined with epithelium across vitelline membrane | 17 | 76 | NA | NA |
| Mesenchyme recombined with epithelium across 0.4, 0.6 $\mu$ porosity Nucleopore filters | 58 | NA | NA | NA |
| Mesenchyme recombined with backskin epithelium | 67 | NA | NA | NA |

NA = not available.

[a]Presented as percent of organ cultured tissues forming
cartilage.

Source: Based on data in Gumpel-Pinot (1972; 1973; 1980).

Chondrogenesis in mouse fore and hind limb buds also requires
a prior epithelial-mesenchymal interaction, independent of later
epithelial influences on limb polarity (Milaire and Mulnard, 1968).
The wealth of accumulated data on biochemistry, cell biology, and
genetics of embryonic limb buds makes this system an ideal one for
those wishing to begin to analyze a skeletogenic epithelial-mesenchymal interaction. Limb mutants such as chondrodystrophies and
achondroplasias where cartilage differentiation and growth are abnormal, have not been examined at all for abnormalities in these
early epithelial-mesenchymal interactions, despite the great utility
of such studies at later stages (Abbott, 1975).

CARTILAGE OF SOMITIC ORIGIN

   Vertebrae of all embryonic vertebrates arise from the sclerotomal portion of the paired somites. Recent evidence (Cheney and
Lash, 1981) establishes the earlier suggestion that the sclerotome,
but not the dermamyotome, is determined for chondrogenesis <u>before</u>
the sclerotomal mesenchyme undergoes its interaction with epithelium. In order to undergo this interaction sclerotomal cells break
away from the remainder of the somite and migrate toward the notochord and ventral half of the spinal cord where they accumulate
(Fig. 9.1). At the same time as sclerotome is migrating, notochord
and spinal cord are synthesizing and depositing extracellular matrices
consisting of microfibrils of "cartilage type" collagen and granules
of "cartilage type" glycosaminoglycans (Linsenmayer et al., 1973;
Frederickson et al., 1977; Carlson and Kenney, 1980; Vasan, 1981).
Sclerotomal cells migrate into and interact with this extracellular
matrix. Evidence for the interaction is of several types. Somites
will not chondrify unless exposed to intact notochord, spinal cord,
or conditioned medium (Hall, 1977). Removal of the extracellular
matrix from notochords prevents them from allowing sclerotomal
mesenchyme to form cartilage (Kosher and Lash, 1975). Notochords

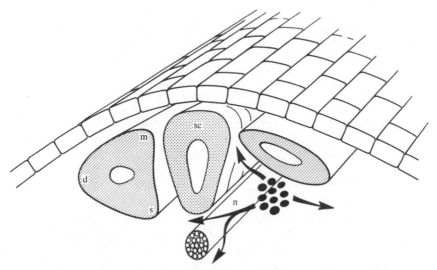

Figure 9.1 A cross section through an embryonic chick to show, on
the left, a somite at H.H. stage 10 (before sclerotomal migration)
and on the right, sclerotomal cells migrating toward the spinal cord
(sc) and notochord (n) at H.H. stage 14 (d, dermatome; m, myotome;
s, sclerotome). Reproduced from Hall (1978a) with permission of
the publisher.

permitted to redeposit extracellular matrix regain inductive ability. It seems clear that this epithelial-mesenchymal interaction is mediated by epithelial extracellular products. As collagen, glycosaminoglycans and/or proteoglycans are the chief extracellular products. Attempts have been made to substitute them, individually, for intact extracellular matrix. Sclerotomal cells will respond to chondroitin-4-sulfate, chondroitin-6-sulfate, cartilage type (type II) collagen, or proteoglycan by heightened synthesis of extracellular matrix products (Kosher et al., 1973; Kosher and Church, 1975; Lash and Vasan, 1978) but cytodifferentiation of cartilage from sclerotomal cells not already engaged in synthesis of extracellular matrix has not yet been achieved. Such positive feedback of cartilage matrix has been seen in various noninductive situations (see Hall, 1978a, pp. 152-155). The implication is that no single component of the notochord or spinal cord extracellular matrices contains the inductive activity. Individual components can augment aspects of expression of the chondrogenic phenotype but a higher organization of extracellular matrix is required for differentiation to be inititiated.

## CARTILAGES OF NEURAL CREST ORIGIN

### Scleral Cartilage

The ring of scleral cartilage which encircles the eyes of birds arises only after mesenchyme has interacted with extracellular matrix products derived from pigmented retinal epithelium. Periocular mesenchyme from H.H. stage 17 or 18 embryos fails to chondrify if maintained alone but does chondrify when grafted in contact with the epithelium (Newsome, 1972; Stewart and McCallion, 1975). Newsome (1976) provided a very convincing demonstration that extracellular products produced by the epithelium were sufficient to elicit chondrogenesis. Pigmented retinal epithelia were cultured on Millipore filters for 4-8 weeks, during which time extracellular products were deposited into the filters. Distilled water lysis was used to remove the epithelial cells before periocular mesenchyme or unmigrated neural crest cells were cultured in contact with the matrix products. Of such cultures, 39 percent formed cartilage. Knowing that pigmented retinal epithelium synthesizes type II ("cartilage type") collagen, Newsome cultured mesenchyme on collagen gels, but did not observe any cartilage formation. Like notochordal extracellular matrix which also deposits type II collagen, a single matrix component cannot substitute for the inductively active extracellular matrix.

The other neural crest-derived cartilages for which there is some information on the mechanism of the epithelial-mesenchymal interaction include the otic capsule and Meckel's cartilage (for a review see Hall, 1982).

## Otic Capsule

Once the epithelial otic vesicle has been induced to form by inductions from both head mesoderm and hind brain (Yntema, 1955; Van de Water et al., 1980) it in turn induces adjacent mesenchyme to form an otic capsule in trout (Benoit, 1957), turtles (Toerien, 1965), frogs and salamanders (Horstadius, 1950; Yntema, 1955; Holtfreter, 1968; Hall, 1982), birds (Benoit and Schowing, 1970; Jaskoll, 1980; Hall, 1982), and mammals (Toerien, 1969; Pugin, 1972). These are not species-specific interactions, since frog and mammalian otic vesicle will induce salamander and chick otic mesenchyme to chondrify (Lewis, 1907; Toerien, 1969; Pugin, 1972). Requirement for otic epithelium is specific in amphibians but not in birds where notochord and spinal cord can be substituted (Benoit, 1956; see also Hall, 1982). Other known chondrogenic mesenchyme in the head and trunk cannot respond to inductive influences from the otic epithelium by initiating cartilage formation (Kaan, 1930).

Little attempt appears to have been made to investigate the mode of action of these interactions. Benoit (1960) showed that saline extracts of otic vesicles contained inductive activity but nothing further is known.

Two tissue interactions where cartilage from the otic capsule acts as the inducer merit a mention. Both involve cartilage induced dedifferentiation in an adjacent tissue. Annular tympanic cartilage in frogs induces overlying epithelium to dedifferentiate and form a tympanic membrane (Helff, 1940), and the avian columella induces some otic capsular chondrocytes to dedifferentiate so as to form the annular ligament which anchors the columella to the capsule (Jaskoll and Maderson, 1978).

## Meckel's Cartilage

There is also information available on the epithelial-mesenchymal interactions which lead to formation of Meckel's cartilage in the lower jaw. Horstadius (1950), Holtfreter (1968), and Hall (1980a; 1982) provide reviews.

In amphibians, Meckel's cartilage is induced to form by mesenchyme coming into contact with pharyngeal endoderm, presumably as it migrates to form the mandibular arch (Cusimano-Carollo, 1963; Drews et al., 1972; Corsin, 1975; Epperlein and Lehmann, 1975). This chondrogenic induction can be passed from one mesenchymal cell to another through cell to cell contact (Holtfreter, 1968; Epperlein and Lehmann, 1975; Minuth and Gruntz, 1980). As with avian otic vesicle, extracts of pharyngeal endoderm contain inductive activity. Notochord can also substitute for pharynx, at least in Ambystoma mexicanum (Holtfreter, 1968). Once induced, Meckel's cartilage may play a role in subsequent induction of teeth, larval beak, and indeed the whole mouth cavity (Cusimano-Carollo, 1963).

In birds cranial neural crest cells can form cartilage before they migrate away from the neural tube (Hall and Tremaine, 1979) but such neural crest cells must first interact with an epithelium (Newsome, 1972; Bee and Thorogood, 1980).

An epithelial-mesenchymal interaction is also required for Meckel's cartilage to form in the mouse but it is mandibular epithelium, encountered by neural crest-derived mesenchyme after migration, which is inductively active (Hall, 1980b). The epithelium becomes active at 10 days of gestation and can be replaced by mandibular epithelium from the embryonic chick. No further information is yet available on this interaction.

## ECTOPIC CARTILAGE INDUCED BY EPITHELIA

What follows is a brief summary of an example of epithelium-inducted cartilage formation which occurs in the adult. Ostrowski and Wlodarski (1971) and Anderson (1976) have reviewed their extensive work on this system, in which a variety of cultured amniotic or epithelial cell lines induce chondrogenesis when injected intramuscularly into cortison-sensitized mice. The responding cells are fibroblasts within connective tissue sheaths of muscle. Fibroblasts located subcutaneously do not chondrify in response to this stimulus (Hancox and Wlodarski, 1972). The inductively active cell lines are heteroploid and many, in contrast to noninductively active cell lines, which are agglutinate with concanavalin A (Wlodarski et al., 1974). This, coupled with the failure of these interactions to occur across filter barriers (Anderson, 1976), implicates direct contact or very close apposition between the membranes of the inducing and responding cells.

## TERATOCARCINOMAS

The cartilages which form in the above animals form in localized tumors which arise after the injection of the epithelial cells. Cartilage also forms in teratocarcinomas which arise after primordial germ cells or germ layers from mammalian embryos are grafted ectopically (Levak-Švajger and Švajger, 1974; Illmensee and Stevens, 1979). Because several of the tissues which form in teratocarcinomas (cartilage, bone, hair, teeth) normally arise as a result of epithelial-mesenchymal interactions, one wonders whether such interactions also occur within developing teratocarcinomas. Bennett et al. (1977) have argued that the very fact that such tissues form so readily within teratocarcinomas can be used as evidence that their inductive requirements are not highly specific. The use of teratocarcinomas to explore epithelial-mesenchymal interactions is wide open—I know of no published studies in this area.

BONES OF THE SKULL

There is a limited amount of information available on induction of dermal bones of the skull. In a series of papers published during the 1960s, and summarized by Benoit and Schowing (1970), Schowing extirpated the notochord or various regions of the brain from chick embryos at H.H. stages 10 and 11 and found that, subsequently, adjacent regions of the skull failed to develop. Experiments involving reversal of the developing brain were also followed by

TABLE 9.2

Epithelial-Mesenchymal Interactions and
Murine Frontal Bone Development[a]

| | Theiler Stage[b] of Tissue Isolation | |
|---|---|---|
| Tissue Combination | 16 | 18 |
| Intact (brain, mesenchyme, epithelium) | 0 | 67 |
| Isolated mesenchyme | 0 | 18 |
| Mesenchyme recombined with epithelium | 25 | 36 |
| Mesenchyme recombined with forebrain | 0 | 0 |

[a]Presented as percent of chorioallantoic grafts forming bone.
[b]Theiler (1972) stages where stage 16 is 10 days of gestation and stage 18 is 11 days of gestation.
Source: Compiled by author.

deletions, implicating site specificity in these interactions. Schow-ing's experiments did not implicate the overlying epithelium in these osteogenic interactions. However, more recently, Tyler (1980) and Hall (1980a) have shown that mesenchyme from the frontal regions of embryonic chicks and mice requires the adjacent epithelium if bone is to form. In our study in the mouse, brain was not required (Table 9.2). Considerably more experiments will have to be done to reconcile these results with Schowing's and to further explore this system.

## SCLERAL BONES

The scleral bones which encircle the eyes of birds are induced to form by interaction with epithelial papillae which form in the sclera (Fig. 9.2a and 9.2b). Blanck et al. (1981) and Fyfe and Hall (1981) have provided scanning electron microscopic analyses of various stages of papilla development.

Several lines of evidence summarized in Hall (1978a) and in Fyfe (1980) show that these epithelial papillae interact with subjacent mesenchyme to allow bone to form. The number of papillae equals the number of bones which will form in both normal and in mutant (scaleless) embryos (Murray, 1943; Palmoski and Goetinck, 1970; Coulombre and Coulombre, 1973). Extirpation of a single papilla, either surgically or pharmacologically, results in absence of the corresponding bone (Coulombre et al., 1962; Johnson, 1973). Isolated

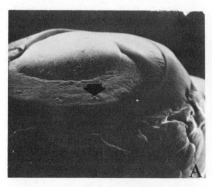

Figure 9.2A The eye from an H.H. stage 33 (7.5 day) embryo to show four of the scleral papillae (arrow) in the scleral (conjunctival) epithelium encircling the lens (L). (x 9)

9.2B A single scleral papilla from an H.H. stage 34 (8-day) embryo. (x 250)

TABLE 9.3

Epithelial-Mesenchymal Interactions and the Development
of Scleral Bones[a]

|  | H.H. Stage of Tissue Isolation | | | |
| Tissue Combination | 30 | 32 | 35 | 36 |
| --- | --- | --- | --- | --- |
| Scleral mesenchyme | 0 | 0 | 0 | 57 |
| Scleral mesenchyme recombined with scleral epithelium | 0 | NA | 33 | 50 |
| Scleral mesenchyme combined with H.H. stage 22 mandibular epithelium | 44 | 67 | 44 | 55 |
| H.H. stage 22 mandibular mesenchyme combined with scleral epithelium | 43 | 83 | 100 | NA |

NA = not available.
[a]Presented as percent of chorioallantoic grafts forming bone.
Source: Based on data from Hall (1981d).

scleral mesenchyme fails to form bone but mesenchyme grafted in
contact with scleral or mandibular epithelium forms bony scleral
ossicles (Table 9.3; see also Hall, 1981d). The highly specialized
morphological changes undergone by papillae may not relate to their
osteogenic inductive activity for they occur after induction has been
initiated (Fyfe and Hall, 1981; Hall, 1981d).

## MEMBRANE BONES OF THE FACIAL SKELETON

In addition to studies showing that epithelial-mesenchymal
interactions initiate development of frontal and scleral bones, tissue
separation and recombination studies have shown that membrane
bones of mandibular, maxillary, and palatine skeletons also arise
in response to such interactions (Tyler and Hall, 1977; Hall, 1978a;
1980b; Tyler, 1981; Tyler and McCobb, 1980a; 1981). Heterotypic
and heterospecific tissue recombinations have shown that these interactions are neither site nor species specific (Hall, 1978b; 1980b;
1981c; Tyler and McCobb, 1980b). As summarized in Table 9.4,

TABLE 9.4

Epithelial-Mesenchymal Interactions and Avian Mandibular and
Maxillary Membrane Bone Development[a]

| Epithelium | Mandibular Mesenchyme[b] | Maxillary Mesenchyme[b] |
|---|---|---|
| | 0 | 0 |
| Mandibular[b] | 84 | 75 |
| Maxillary[b] | 73 | 62 |
| Scleral (H.H. stages 31, 32) | 62 | NA |
| Scleral (H.H. stage 35) | 100 | NA |
| Leg bud (H.H. stages 18–24) | 63 | 47 |
| Wing bud (H.H. stages 24–28) | 58 | 50 |
| Back (H.H. stage 31) | 57 | NA |
| Abdominal (H.H. stage 31) | 66 | NA |
| Mouse mandibular (10 days of gestation) | 71 | NA |

NA = not available.
[a]Presented as percent of chorioallantoic grafts forming bone.
[b]From H.H. stage 22 embryos.
Source: Based on data from Hall (1981c).

epithelia from sites where membrane bone does not form (back,
abdomen)* have the capacity to initiate membrane bone formation
from mandibular mesenchyme. This nonspecificity fulfills one of
the several criteria enumerated by Wessells (1977) as distinguishing
a permissive from an instructive tissue interaction. A second criterion for a permissive interaction is inability of the inducer (epithelium) to elicit differentiation (osteogenesis) from a site where
the tissue, in this case membrane bone, would not normally occur.
A series of heterotypic and heterospecific tissue recombinations
has established that mandibular and maxillary epithelia only elicit
osteogenesis from mesenchyme which would normally ossify. Thus

---

*I have not included wing and leg buds for membrane bone
does form subperiosteally around the shafts of their long bones,
comprising some 75% of the bulk of long bone in the embryonic chick.

maxillary, scleral, and murine mandibular mesenchyme, which
normally form membrane bone, respond to mandibular epithelia by
forming bone, but wing, back, chorioallantoic, or trunk neural
crest-derived mesenchyme which do not form membrane bone in ovo,
did not form bone when combined with mandibular epithelium (Table
9.5). Clearly, the specificity resides within local populations of
osteogenic mesenchyme with epithelia providing a permissive, but
absolutely essential, environment.

Even though other epithelia can substitute for epithelium of
the mandibular arch, there is a temporal specificity to these interactions. Mandibular epithelia lose their ability to induce at H.H.
stage 24, coincident with mesenchyme becoming independent of any
further requirement for an epithelial interaction (Hall, 1978b). Although wing and leg bud epithelia are both inductively active, epithelium of leg bud is active from at least as early as H.H. stage 18,
but wing bud epithelium does not acquire inductive activity until
H.H. stage 24 (Hall, 1978b). Similarly, in the mouse, mandibular

TABLE 9.5

Epithelial-Mesenchymal Interactions Involving
Mandibular Epithelium Are Permissive[a]

| Source of Mesenchyme | Percent of Grafts with Bone |
|---|---|
| Mandible[b] | 84 |
| Maxilla[b] | 75 |
| Sclera[c] | 42 |
| Mouse mandible[d] | 45 |
| Wing bud[b] | 0 |
| Back[b] | 0 |
| Chorioallantois | 0 |
| Trunk neural crest[e] | 0 |

[a] Each experiment involved recombining H.H. stage
22 mandibular epithelium with one of the mesenchymes
and grafting to the chorioallantoic membrane.
[b] H.H. stage 22.
[c] H.H. stages 30-32.
[d] Theiler stage 14 (9 days) of gestation.
[e] H.H. stage 10.
Source: Based on data from Hall (1981c).

TABLE 9.6

Involvement of Epithelial Collagen in Induction of
Avian Mandibular Membrane Bone

| Inhibitor | Site of Blocking Action | Osteogenesis[a] |
|---|---|---|
| None | | + |
| BUdR | Cell differentiation | 0 |
| L-azetidine-2-carboxylic acid | Proline hydroxylation | 0 |
| $\alpha, \alpha'$-Dipyridyl | Proline hydroxylation | 0 |
| Penicillamine | Collagen polymerization | + |
| 4-Methylumbelliferone | Glycosaminoglycan synthesis | + |
| Streptomyces hyaluronidase | Hyaluronate synthesis | + |
| Testicular hyaluronidase | Chondroitin sulfate synthesis | + |

[a]+, bone present; 0, bone absent.
Source: Based on the recombination of inhibited epithelia with untreated mesenchyme. Data in part from Bradamante and Hall (1980) and in part from Hall (unpublished).

mesenchyme from 9 day-old embryos is able to respond to epithelia but mandibular epithelium is not inductively active until 10 days of gestation (Hall, 1980b).

Once we have a clearer idea of the mechanisms underlying the above interactions we will be able to begin to explore the basis for these temporal differences in inductive activity. A start has been made on this problem in my laboratory. In addition to the study identifying temporal aspects of epithelial inductive activity (Hall, 1978b), we have shown that induction is correlated with proliferative activity of the epithelium (Hall, 1980c). Epithelia which induce ectopic bone or cartilage to form are also only inductively active when dividing (Huggins, 1931; Wlodarski, 1969). Whether mitotic activity bears a direct relationship to induction (e.g., epithelial cells only inductive at a particular phase of the cell cycle) or is only a correlate (e.g., the more epithelial cells present, the more active the epithelium) remains to be determined.

We do now have considerable direct and indirect evidence indicating that epithelial inductive activity resides in extracellular

TABLE 9.7

Formation of Bone in Response to Epithelial Basal
Lamina Deposited in Vitro[a]

| Treatment | Percent with Bone |
|---|---|
| Mesenchyme cultivated on epithelial cells | 48 |
| Mesenchyme cultivated on: | |
| 1. Epithelial extra-cellular products (e.e.p.) | 75 |
| 2. Testicular hyaluronidase-treated e.e.p. | 37 |
| 3. Trypsin-treated e.e.p. | 0 |
| 4. L-azetidine-2-carboxylic acid-treated e.e.p. | 0 |

[a]Mandibular mesenchyme was grown in association
with epithelial cells or with the extracellular products
which they deposited after 28 days in vitro.
Source: Based on data in Hall and Van Exan (1981).

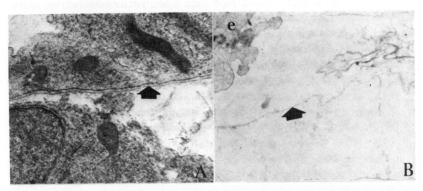

Figure 9.3A  A continuous basal lamina (arrow) separates epithelium
(above) from mesenchyme (below) in the mandibular arch of an H.H.
stage 22 embryonic chick. Mesenchymal cell processes contact the
basal lamina. (x 13,300)

9.3B  Mandibular epithelia (e) cultured for 28 days deposit a
basal lamina-like (arrow) osteogenically inductive, extracellular
matrix. Mesenchyme cultivated on this material forms bone. For
details see Hall and Van Exan (1981). (x 13,300)

TABLE 9.8

Transfilter Recombinations of Mandibular Epithelium
and Mesenchyme and the Initiation of Osteogenesis[a]

| Filter Porosity ($\mu$) | Filter Thickness ($\mu$) | Percent of Grafts with Bone |
|---|---|---|
| Millipore Filters | | |
| 0.45 | 125 | 0 |
| 0.45 | 25 | 0 |
| Nucleopore Filters | | |
| 0.8 | 10 | 87 |
| 0.4 | 10 | 82 |
| 0.1 | 5 | 37 |
| 0.03 | 5 | 0 |

[a]Transfilter recombinations were grafted to the
chorioallantoic membranes of host embryos.
Source: Based on data in Van Exan and Hall
(1982).

products located within the basal lamina. Pharmacological inhibition
of synthesis of epithelial collagen followed by recombination of inhibited epithelia with untreated mandibular mesenchyme prevents that
mesenchyme from forming bone (Bradamante and Hall, 1980;
Table 9.6). Inhibition of collagen polymerization or of epithelial
glycosaminoglycan synthesis does not render epithelia inactive
(Table 9.6). The clear implication of this indirect evidence is that
epithelially-derived collagen is required for mandibular epithelium
to induce mandibular mesenchyme to form bone.

In a more direct test we have shown, using the same experimental design as used by Newsome (1976) in his study on the induction
of scleral cartilage, that mesenchyme can form bone in response to
epithelial products deposited onto Millipore filters (Hall and Van
Exan, submitted). Trypsin, L-azetidine-2-carboxylic acid (LACA),
but not hyaluronidase treatment, removed the inductive activity,
which was shown, using transmission electron microscopy, to
reside in basal lamina-like material deposited onto the filters
(Table 9.7; Fig. 9.3a and 9.3b). Such material, along with inductive

activity, was removed by LACA treatment (Hall and Van Exan, submitted). These observations, along with the finding that (1) more mesenchymal cell processes contact inductively active epithelial basal lamina than contact the noninductive portion of the mandibular epithelium (Hall and Van Exan, submitted); and (2) that this interaction does not occur across Millipore filters or across Nucleopore filters with a porosity of less than $0.1 \mu$ (Van Exan and Hall, submitted; Table 9.8), indicate that mandibular mesenchyme must make contact with the collagen within the basal lamina for the epithelial-mesenchymal interaction to be initiated.

## CONCLUSIONS

As I emphasized in the beginning of this chapter, I believe that it is dangerous to attempt to generalize the results presented here. This caution is well illustrated by the initiation of Meckel's cartilage in amphibians, birds, and mammals. In all three groups an epithelial-mesenchymal interaction governs the initiation of chondrogenesis. So far, we can safely generalize. However, in birds the interaction occurs before neural crest cells migrate away from the mesencephalic neural tube. In amphibians, the interaction occurs during, and in mammals it occurs after migration. Timing of these interactions cannot be generalized, nor can behavior of one group be predicted from that observed in another. Even though timing differs, and even though different epithelia are involved in each group (cranial ectoderm in birds, pharyngeal endoderm in amphibians, mandibular epithelium in mammals—although even here I generalize for the mouse is the only mammal which has been examined), the basic mechanism of the interaction may be the same in all three groups. We just do not know. In amphibians, the induction can be passed from one mesenchymal cell to another via cell contact but homogenates of pharyngeal endoderm contain inductive activity (Holtfreter, 1968). This important question of whether every mesenchymal cell has to contact epithelium, or whether, once induced, mesenchymal cells can establish a chain of induction, has not been addressed in any other skeletogenic system.

It is becoming clear that epithelial chondrogenic or osteogenic inductive activity usually resides within an extracellular matrix deposited by the epithelium, and that single extracellular matrix products cannot substitute for intact extracellular matrix. Extracellular localization of epithelial inductive activity has been shown for limb cartilage (basement membrane of epithelium), somitic cartilage (notochordal and spinal cord extracellular matrices), scleral cartilage (matrix deposited by pigmented retinal epithelium), and for mandibular membrane bones (basal lamina of mandibular epithelium). We do not know whether it is the extracellular fraction of the cell homogenates which induces otic capsule and Meckel's cartilage to chondrify. In any event it is clear that we need to focus our attention on (1) the constitution and organization of epithelial basal laminae; (2) the contact of mesenchymal cells with basal laminae; and (3) on ways of isolating intact, inductively active basal laminae and extracellular matrices. As indicated in other chapters in this volume, knowledge of the biochemistry of basal laminae is increasing rapidly. The presence of fibronectin, laminin, and various collagens and glycoproteins in amounts which vary from place to place and time to time provides epithelia with tremendous potential for mediating inductive interactions. We (Hall and Van Exan, submitted) have described the complexity which arises when the three factors of differential distribution of competent mesenchyme and of inductively active epithelia, and varying amounts of contact between epithelium and mesenchyme are all operating in controlling a single epithelial-mesenchymal interaction. Clearly we shall need to utilize a variety of approaches in each system if we are to unravel the complexity which seems to underlie each such interaction. Given that all cartilages and bones in embryos and many of those which arise ectopically or in tumors in adults, arise because of prior tissue interactions, understanding their bases is a major challenge for this and subsequent generations of developmental biologists.

## ACKNOWLEDGEMENTS

The author's original research has been supported by the Natural Sciences and Engineering Research Council of Canada, by the Research Development Fund in the Sciences of Dalhousie University and by the Nuffield Foundation.

## REFERENCES

Abbott, U. K. (1975). Genetic approaches to studies of tissue interactions. Genet. Lect. 4:69-84.

Anderson, H. C. (1976). Osteogenetic epithelial-mesenchymal cell interactions. Clin. Orthop. 119:211-224.

Bee, J. and Thorogood, P. V. (1980). The role of tissue interactions in the skeletogenic differentiation of avian neural crest cells. Dev. Biol. 78:47-62.

Bennett, D., Artzt, K., Magnuson, T., and Spiegelman, M. (1977). Developmental interactions studied with experimental teratomas derived from mutants at the T/t locus in the mouse. In Cell Interactions in Differentiation (M. Karkinen-Jääskeläinen, L. Saxen, and L. Weiss, eds.), pp. 389-398. Academic Press, New York.

Benoit, J. A. A. (1956). Chondrogenèse otique après excision de l'otocyste et implantation de moelle ou de chord chez l'embryon de poulet. C. R. Soc. Biol. (Paris) 150:240-242.

Benoit, J. A. A. (1957). Irradiation localisée de l'otocyste chez les embryons de poulet et de truite. Bull. Soc. Zool. France 82: 238-243.

Benoit, J. A. A. (1960). Induction de cartilage in vitro par l'extrait d'otocystes d'embryons de poulet. J. Embryol. Exp. Morphol. 8:33-38.

Benoit, J. A. A. and Schowing, J. (1970). Morphogenesis of the neurocranium. In Tissue Interactions during Organogenesis (E. Wolff, ed.), pp. 105-130. Gordon Breach, New York.

Bergsma, D. and Langmen, I., eds. (1975). Morphogenesis and Malformation of Face and Brain, Birth Defects: Original Article Series. Vol. XI, no. 7. Alan R. Liss, New York.

Blanck, C. E., McAleese, S. R., and Sawyer, R. H. (1981). Morphogenesis of conjunctival papillae from normal and scaleless chick embryos. Anat. Rec. 199:249-257.

Bradamante, Z. and Hall, B. K. (1980). The role of epithelial collagen and proteoglycan in the initiation of osteogenesis by avian neural crest cells. Anat. Rec. 197:305-315.

Carlson, E. C. and Kenney, M. C. (1980). Surface ultrastructure of the isolated avian notochord in vitro: the effect of the perinotochordal sheath. Anat. Rec. 197:257-276.

Cheney, C. M. and Lash, J. W. (1981). Diversification within embryonic chick somites: differential response to notochord. Dev. Biol. 81:288-298.

Corsin, J. (1975). Influence du hyaluronate et de la hyaluronidase sur la chondrogenese cephalique chez les amphibiens. Acta Embryol. Exp. 1:15-22.

Coulombre, A. J. and Coulombre, J. L. (1973). The skeleton of the eye. II. Overlap of the scleral ossicles of the domestic fowl. Dev. Biol. 33:257-267.

Coulombre, A. J., Coulombre, J. L., and Mehta, H. (1962). The skeleton of the eye. I. Conjunctival papillae and scleral ossicles. Dev. Biol. 5:382-401.

Cusimano-Carollo, T. (1963). Investigation on the ability of the neural folds to induce a mouth in the Discoglossus pictus embryos. Acta Embryol. Morphol. Exp. 6:158-169.

Drews, U., Kocher-Becker, U., and Drews, U. (1972). The induction of visceral cartilage from cranial neural crest by pharyngeal endoderm in hanging drop cultures and the locomotory behaviour of the neural crest cells during cartilage differentiation. Wilhelm Roux's Archives 171:17-37.

Epperlein, H. H. and Lehmann, R. (1975). Ectomesenchymal-endodermal interaction system (EEIS) of Triturus alpestris in tissue culture. 2. Observations on differentiation of visceral cartilage. Differentiation 4:159-174.

Frederickson, R. G., Morse, D. E., and Low, F. N. (1977). High-voltage electron microscopy of extracellular fibrillogenesis. Am. J. Anat. 150:1-34.

Fyfe, D. MacG. (1980). The morphogenesis of scleral papillae and scleral ossicles in the eye of the chick embryo. M.Sc. Thesis, Dalhousie University, Halifax.

Fyfe, D. MacG. and Hall, B. K. (1981). A scanning electron microscopic study of the developing epithelial scleral papillae in the eye of the embryonic chick. J. Morphol. 167:201-209.

Gorlin, R. J., ed. (1980). Morphogenesis and Malformation of the Ear, Birth Defects: Original Article Series. Vol. XVI, no. 4. Alan R. Liss, New York.

Gumpel-Pinot, M. (1972). Culture in vitro de l'ébauche de l'aile de l'embryon de poulet. Rôle de l'ectoderme dur la chondrogenèse. C. R. Acad. Sci. (Paris) 274:2786-2789.

Gumpel-Pinot, M. (1973). Culture in vitro du bourgeon d'aile de l'embryon de poulet: différenciation du cartilage. Ann. Biol. 12:417-429.

Gumpel-Pinot, M. (1980). Ectoderm and mesoderm interactions in the limb bud of the chick embryo studies by transfilter cultures; cartilage differentiation and ultrastructural observations. J. Embryol. Exp. Morphol. 59:157-173.

Gumpel-Pinot, M. (1981). Ecto-mesodermal interactions and chick embryo limb chondrogenesis. Ultrastructural studies on cultures in the vitelline membrane. Arch. d'Anat. Microsc. Morphol. Exp. 70:1-14.

Hall, B. K. (1977). Chondrogenesis of the somitic mesoderm. Adv. Anat. Embryol. Cell Biol. 53:1-50.

Hall, B. K. (1978a). Developmental and Cellular Skeletal Biology. Academic Press, New York.

Hall, B. K. (1978b). Initiation of osteogenesis by mandibular mesenchyme of the embryonic chick in response to mandibular and non-mandibular epithelia. Arch. Oral Biol. 23:1157-1161.

Hall, B. K. (1980a). Chondrogenesis and osteogenesis in cranial neural crest cells. In Current Research Trends in Prenatal Craniofacial Development (R. M. Pratt and R. L. Christiansen, eds.), pp. 47-63. Elsevier/North-Holland, New York.

Hall, B. K. (1980b). Tissue interactions and the initiation of osteogenesis and chondrogenesis in the neural crest-derived mandibular skeleton of the embryonic mouse as seen in isolated murine tissues and in recombinations of murine and avian tissues. J. Embryol. Exp. Morphol. 58:251-264.

Hall, B. K. (1980c). Viability and proliferation of epithelia and the initiation of osteogenesis within mandibular ectomesenchyme in the embryonic chick. J. Embryol. Exp. Morphol. 56:71-89.

Hall, B. K. (1981a). Intra- and extracellular control of the differentiation of cartilage and bone. Histochem. J. (in press).

Hall, B. K. (1981b). Embryogenesis:—cell-tissue interactions. In Skeletal Research—An Experimental Approach (D. J. Simmons and A. S. Kunin, eds.), Vol. 2 (in press). Academic Press, New York.

Hall, B. K. (1981c). The induction of neural crest-derived cartilage and bone by embryonic epithelia: an analysis of the mode of action of an epithelial-mesenchymal interaction. J. Embryol. Exp. Morphol. (in press).

Hall, B. K. (1981d). Specificity in the differentiation and morphogenesis of neural crest-derived scleral ossicles and of epithelial scleral papillae in the eye of the embryonic chick. J. Embryol. Exp. Morphol. (in press).

Hall, B. K. (1982). Tissue interactions and chondrogenesis. In Cartilage (B. K. Hall, ed.), Vol. 2 (in press). Academic Press, New York.

Hall, B. K. and Tremaine, R. (1979). Ability of neural crest cells from the embryonic chick to differentiate into cartilage before their migration away from the neural tube. Anat. Rec. 194:469-476.

Hall, B. K. and Van Exan, R. J. (1981). Induction of bone by epithelial cell products. (submitted)

Hall, B. K., and Van Exan, R. J. (1982). Distribution of osteo- and chondrogenic neural crest cells and of osteogenically inductive epithelia in mandibular arches of embryonic chicks. (submitted)

Hancox, N. M. and Wlodarski, K. (1972). The role of host site in bone induction by transplanted xenogenic epithelial cells. Calcif. Tissue Res. 8:258-261.

Helff, O. M. (1940). Studies on amphibian metamorphosis. XVII. Influence of non-living annular tympanic cartilage on tympanic membrane formation. J. Exp. Biol. 17:45-60.

Hinchliffe, J. R. and Johnson, D. R. (1980). The Development of the Vertebrate Limb. Oxford University Press, Oxford.

Holtfreter, J. (1968). Mesenchyme and epithelia in inductive and morphogenetic processes. In Epithelial-Mesenchymal Interactions (R. Fleischmajer and R. E. Billingham, eds.), pp. 1-30. Williams & Wilkins, Baltimore.

Horstadius, S. (1950). The Neural Crest: its Properties and Derivatives in the Light of Experimental Research. Oxford University Press, Oxford.

Huggins, C. B. (1931). The formation of bone under the influence of epithelium of the urinary tract. Arch. Surg. 22:377-408.

Illmensee, K. and Stevens, L. C. (1979). Teratomas and Chimeras. Sci. Am. 240:121-132.

Jaskoll, T. F. (1980). Morphogenesis and teratogenesis of the middle ear in animals. In Genetic and Environmental Hearing Loss: Syndromic and Nonsyndromic (L. S. Levin and C. H. Knights, ed.), Birth Defects: Original Article Series. Vol. XVI, no. 7. Alan R. Liss, New York.

Jaskoll, T. F. and Maderson, P. F. A. (1978). A histological study of the development of the avian middle ear and tympanum. Anat. Rec. 190:177-200.

Johnson, L. G. (1973). Development of chick embryo conjunctival papillae and scleral ossicles after hydrocortisone treatment. Dev. Biol. 30:223-227.

Kaan, H. W. (1930). The relation of the developing auditory vesicle to the formation of the cartilage capsule in Ambystoma punctatum. J. Exp. Zool. 55:263-291.

Kosher, R. A. and Church, R. L. (1975). Stimulation of in vitro somite chondrogenesis by procollagen and collagen. Nature 258: 327-329.

Kosher, R. A. and Lash, J. W. (1975). Notochordal stimulation of in vitro somite chondrogenesis before and after enzymatic removal of perinotochordal materials. Dev. Biol. 42:362-378.

Kosher, R. A., Lash, J. W., and Minor, R. R. (1973). Environmental enhancement of in vitro chondrogenesis. IV. Stimulation of somite chondrogenesis by exogenous chondromucoprotein. Dev. Biol. 35:210-220.

Lash, J. W. and Vasan, N. S. (1978). Somite chondrogenesis in vitro. Stimulation by exogenous extracellular matrix components. Dev. Biol. 66:151-171.

Levak-Švajger, B. and Švajger, A. (1974). Investigation on the origin of the definitive endoderm in the rat embryo. J. Embryol. Exp. Morphol. 32:445-459.

Lewis, W. H. (1907). On the origin and differentiation of the otic vesicle in amphibian embryos. Anat. Rec. 1:141-145.

Linsenmayer, T. F., Trelstad, R. L., and Gross, J. (1973). The collagen of chick embryonic notochord. Biochem. Biophys. Res. Commun. 53:39-45.

Melnick, M. and Jorgenson, R., eds. (1979). Developmental Aspects of Craniofacial Dysmorphology, Birth Defects: Original Article Series. Vol. XV, no. 8. Alan R. Liss, New York.

Milaire, J. and Mulnard, J. (1968). Le rôle de l'epiblaste dans la chondrogénèse du bourgeon de membre chez la souris. J. Embryol. Exp. Morphol. 20:215-236.

Minuth, M. and Gruntz, H. (1980). The formation of mesodermal derivatives after induction with vegetalizing factor depends on secondary cell interactions. Cell Differ. 9:229-238.

Morriss, G. M. and Thorogood, P. V. (1978). An approach to cranial neural crest cell migration and differentiation in mammalian embryos. In Development of Mammals (M. H. Johnson, ed.), pp. 363-412. Elsevier/North-Holland, Amsterdam.

Murray, P. D. F. (1943). The development of the conjunctival papillae and of the scleral bones in the chick embryo. J. Anat. 77: 225-240.

Newsome, D. A. (1972). Cartilage induction by retinal pigmented epithelium of chick embryos. Dev. Biol. 27:575-579.

Newsome, D. A. (1976). In vitro stimulation of cartilage in embryonic chick neural crest cells by products of retinal pigmented epithelium. Dev. Biol. 49:496-507.

Ostrowski, K. and Wlodarski, K. (1971). Induction of heterotopic bone formation. In Biochemistry and Physiology of Bone (G. H. Bourne, ed.), pp. 299-337. Academic Press, New York.

Palmoski, M. J. and Goetinck, P. F. (1970). An analysis of the development of conjunctival papillae and scleral ossicles in the eye of the scaleless mutant. J. Exp. Zool. 174:157-164.

Pratt, R. M. and Christiansen, R. L. (1980). Current Research Trends in Prenatal Craniofacial Development. Elsevier/North-Holland, New York.

Pugin, E. (1972). Induction de cartilage après excision de la cupule otique chez l'embryon de poulet, par des greffons d'organes embryonnaires de souris. C. R. Acad. Sci. (Paris) 275:2543-2546.

Scott-Savage, P. and Hall, B. K. (1979). The timing of the onset of osteogenesis in the tibia of the embryonic chick. J. Morphol. 162:453-464.

Scott-Savage, P. and Hall, B. K. (1980). Differentiative ability of the tibial periosteum from the embryonic chick. Acta Anat. 106: 129-140.

Searls, R. L. (1968). Development of the embryonic chick limb bud in avascular culture. Dev. Biol. 17:382-399.

Stewart, P. A. and McCallion, D. J. (1975). Establishment of the scleral cartilage in the chick. Dev. Biol. 46:383-389.

Summerbell, D. (1974). A quantitative analysis of the effect of excision of the AER from the chick limb bud. J. Embryol. Exp. Morphol. 32:651-660.

Theiler, K. (1972). The House Mouse. Development and Normal Stages from Fertilization to 4 Weeks of Age. Springer-Verlag, Berlin.

Toerien, M. J. (1965). An experimental approach to the development of the ear capsule in the turtle, Chelydra serpentina. J. Embryol. Exp. Morphol. 13:141-149.

Toerien, M. J. (1969). Die ontwikkeling van die inwendige oor van die konyn in kuikenkoppe. Tydskr. Natuurw. June-Sept.:152-155.

Tyler, M. S. (1980). Tissue interactions in the development of neural crest-derived membrane bones. Am. Zool. 20:944 (Abstract).

Tyler, M. S. (1981). Reciprocal tissue interactions in the secondary palate of the embryonic chick. J. Dent. Res. 60A:316 (Abstract).

Tyler, M. S. and Hall, B. K. (1977). Epithelial influences on skeletogenesis in the mandible of the embryonic chick. Anat. Rec. 188:229-240.

Tyler, M. S. and McCobb, D. P. (1980a). The genesis of membrane bone in the embryonic chick maxilla: epithelial-mesenchymal tissue recombination studies. J. Embryol. Exp. Morphol. 56: 269-281.

Tyler, M. S. and McCobb, D. P. (1980b). The specificity of epithelial requirements for osteogenesis in the craniofacial region. In Current Research Trends in Prenatal Craniofacial Development (R. M. Pratt and R. L. Christiansen, eds.), p. 450. Elsevier/North-Holland, New York.

Tyler, M. S. and McCobb, D. P. (1981). Tissue interactions promoting osteogenesis in the embryonic chick palate. Arch. Oral Biol. (in press).

Van De Water, T. R., Maderson, P. F. A., and Jaskoll, T. F. (1980). The morphogenesis of the middle and external ear. In Morphogenesis and Malformation of the Ear (R. Gorlin, ed.), pp. 147-180. Alan R. Liss, New York.

Vasan, N. (1981). Analysis of perinotochordal materials. 1. Studies on proteoglycan synthesis. J. Exp. Zool. 215:229-233.

Wessells, N. K. (1977). Tissue Interactions and Development, W. A. Benjamin, Menlo Park.

Wlodarski, K. (1969). The inductive properties of epithelial established cell lines. Exp. Cell Res. 57:446-448.

Wlodarski, K., Ostrowski, K., Chtopkiewicz, B., and Koziorowska, J. (1974). Correlation between the allutinability of living cells by concanavalin A and their ability to induce cartilage and bone formation. Calcif. Tissue Res. 16:251-256.

Yntema, C. L. (1955). Ear and nose. In Analysis of Development (B. H. Willier, P. A. Weiss, and V. Hamburger, eds.), pp. 415-428. W. B. Saunders, Philadelphia.

# 10

# AN EVOLUTIONARY VIEW OF EPITHELIAL-MESENCHYMAL INTERACTIONS
## Paul F. A. Maderson

I became a neontologist rather than a paleontologist since I have always suspected that few paleontologists appreciate that fossils result from developmental processes. The proposal that the mechanisms underlying evolutionary change were associated with tissue interactions (Maderson, 1975) resulted, I was told, from the "systems analysis" approach which I had pursued. At that time I did not know what systems analyses were any more than I knew that I was one of many "closet believers" in a relationship between ontogeny and phylogeny (Gould, 1977, pp. 1-2). My long-held acceptance of such a relationship springs from the influence of instructors during my undergraduate years, especially the late Dr. S. M. Manton, F.R.S. I was taught that an organism exists as an entity, and interacts with its environment, from fertilization through death. One may therefore seek the adaptive significance of any and all phenotypic traits appearing at any stage in the life cycle: where such cannot be readily detected, there are grounds for considering the possibility that one is observing form with possible phyletic significance. If saying this makes me a "closet neo-Haeckelian," so be it; my own feeling has always been that Haeckelians were misguided rather than wrong. It is indisputable that recapitulation of form occurs, what is needed is an appropriate evaluation of its biological role.

It is appropriate that a book on tissue interactions should include a discussion of their evolutionary significance for two very different reasons. First, many of the specific questions that have been addressed here implicitly reflect acceptance of the evolutionary process to such a degree that the questions themselves become de facto, explicitly evolutionary questions. A mechanistic developmental

biologist would never ask, "Do the same rules governing axial determination in the chick wing apply to amphibian legs?" "Can mammalian
ectoderm be induced to form feathers?" "Why don't birds have teeth?"
In fact, all biological research assumes the validity of evolutionary
thinking, else why would someone studying a particular phenomenon
in a bird cite the results of a study dealing with insects? This apparent philosophical digression leads to my second point. The renais-
sance of interest in morphogenesis represents, to a certain degree,
a reaction against the as yet unrealized promise that molecular
studies alone will explain embryogenesis. This quiet revolution has
a noisier counterpart in evolutionary biology. From the late 1930s
through the early 1970s, evolution was assumed to be fully explicable
in terms of the so-called "Synthetic Theory" (Mayr and Provine,
1980; Maderson et al., 1982) which is presently under attack from
two directions. First, the proposal of "punctuated equilibria" as an
alternative to traditional "gradualistic" models of evolutionary change
by Eldredge and Gould (1972) has engendered much controversy (see
refs. in Maderson et al., 1982), but especially important here is
the fact that it has reawakened interest in macrorevolutionary phenomena (Maderson et al., 1982). Second, and this applies to all
forms of evolutionary theory, there is a vocal group which contends
that since such theories cannot be tested directly, they are nonscien-
tific and therefore evolutionary speculation is spurious. This philosophical criticism is of interest here insofar as I hope to demonstrate
that, in principle, it is now possible to plan and execute experiments
using the techniques of developmental biology to examine how
morphological changes which are documented in the fossil record
could have occurred.


THE COMMONALITIES OF LIMBS
AND DENTAL SYSTEMS

    In many biological contexts it is increasingly obvious that
disagreement and polarization of viewpoint spring from semantic
misunderstandings which are to be avoided, especially by a reviewer
attempting an interdisciplinary evaluation. Ideas, hypotheses models,
and concepts have value only insofar as they offer guidelines for
future research and thinking so that their internal consistency and
use of appropriate (and correct) data are of much more value than
the specific arguments. I will write within the confines of a specific
terminology to ensure that I antagonize as few readers as possible.
    I use the phrase "dental system" since discussing teeth alone
in the present context would be as inappropriate (and as impossible)
as sole and specific reference to the femur or humerus. Following

the terminology of Bock and von Wahlert (1965), limbs and dental systems are complex anatomical units comprising many smaller features, e.g., skeletal elements, muscles, integumentary derivatives, etc. For any given tetrapod the entire complexes, and their constituent features have gross recognizable forms. Form (Bock and von Wahlert, 1965) refers to the size, shape, cellular composition, etc. The form of a feature permits many different functions, the properties deriving from the chemical, physical, and cellular aspects of form, e.g., a limb can support weight because of the mechanical strength of its skeletal features, a tooth can cut or grind because its enamel and/or dentine are hard. This functional concept is here extended to include morphogenetic properties which cells may manifest during organogenesis. The faculty is defined as a combination of form and function, thus representing what a feature is <u>capable</u> of doing in the life of an organism. The biological role is what natural selection "sees," albeit indirectly through a monitoring of the organism's total fitness (Maderson, 1975, p. 322). Morphogenetic sequences may be modulated so that the form(s) and function(s) contributing to a particular biological role may be "improved" in the face of a particular selection pressure. Such modulations may enhance, indeed produce, faculties not previously the subject of selection but potentially important to an organism's survival in a changed environment. Within the context of Gans' (1979) thesis of "excessive construction" it would be such modulations of morphogenetic activity which would bring about changes in form and function of a feature so that different faculties could become the primary focus of selection in a changed environment. Thus "new" biological roles—those which selection now "sees"—are but a new emphasis of preexisting faculties, the corollary of which is that "neomorphic" features are but a posteriori identifiable forms of features previously the subject of selection in a different context.

Like all complex anatomical units, limbs and dental systems comprise many features whose forms resemble those of features seen elsewhere in the body (e.g., striated muscles, bone, etc.) but also possess unique features (e.g., claws, exposed mineralizations, etc.). Their biological roles are incompletely understood mechanistically in the sense that we have little knowledge of the precise nature of motor control of muscles and proprioceptive feedback thereof. Thus, when we speak of locomotory or feeding efficiency in extant species, we are making very simple approximations. It follows that when we propose biological roles for the fossil remains of complex anatomical units we must be even more aware that we are arguing from analogy with a relatively narrow data base for modern species.

Limbs and dental systems are fairly well-known as fossils, although intact remains (i.e., complete sets of skeletal features) or complete jaws associated with neurocrania are relatively rare or insufficiently numerous to permit morphometric analysis of variability within so-called "chronospecies" (Stanley, 1979).

The constituent features of limbs and dental systems have been fairly well described embryologically from biochemical through cellular to tissue and organ levels for a number of modern species. However, we should be cautious in generalization since there are still many gaps in our knowledge.

Two major points emerge from data derived from descriptive and experimental studies of the two systems. First, each results from complex interactions between cell populations of diverse origin. Second, normal development can be altered by various manipulations. However, since we are forced to assume that similar cellular dynamics underlie the emergence of form in extinct organisms, we should examine the data carefully to ascertain how many possible alternative explanations might exist. A striking example of the dangers of premature generalization is Cameron and Fallon's (1977) demonstration that amphibian digitogenesis is due to differential mitotic activity, an observation which led to the realization that the much better known involvement of cell death is in fact an amniote characteristic. Such issues must be borne in mind when attempts are made to explain the origin of convergent form in different phyletic lineages. Results from experimental studies may not always be especially valuable in an evolutionary context for two quite different reasons. First, experiments revealing autonomous potentials of cell populations may not have any evolutionary significance: inasmuch as we rightly criticize Haeckelian use of morphological data, dynamic data can be equally misused (see discussion, Maderson, 1967; 1975, p. 317). Second, the data on the origin of dorsoventral and antero-posterior polarity of the limb (see references in Hinchcliffe and Johnson, 1980) are only marginally useful to an evolutionary biologist because our understanding of limb mechanics in living animals is as yet so inadequate. In summary, data contributing to our understanding of the genesis of form represent the major potential contribution of developmental studies to evolutionary biology, but it is important to recognize that such contributions may have usefulness as limited as any from paleontology or comparative anatomy. Nowhere can the limitations of all three disciplines be better illustrated than with reference to the extensive locomotory repertoire and the mothers' ability to transport newly hatched young held in the teeth which have been reported in living crocodilians (Pooley and Gans, 1976); such biological roles could never have been predicted from descriptions of embryonic, adult, or fossil form.

Students who explore tissue interactions are ultimately concerned with pattern formation which, as Wolpert (1981) has emphasized, is different from morphogenesis. Wolpert writes "[While the more limited meaning of morphogenesis] refers to changes in form . . . pattern formation is the process whereby the spatial organization of cellular differentiation is specified, whereas [morphogenesis] involves a process whereby cell sheets change shape or cells move to new positions." (1981, p. 5) Among other issues, the evolutionary biologist is concerned with the how and the why of changes in adult form, i.e., the ultimate expression of morphogenesis within Wolpert's explicit definition. But since the proximate cause of those changes lies in changes in the expression of pattern formation, an embryological phenomenon, its causality cannot be addressed by study of fossils. A functional anatomist need refer to neither developmental nor paleontological data and may elect to ignore—even reject—the evolutionary process. However, developmental data can supply the necessary links between disciplines.

Advances in our knowledge of morphogenetic processes over the past 25 years have revealed, and continue to reveal, the ubiquity of certain cellular phenomena (Maderson, 1975; Wessells, 1982; Lovtrup, 1981). The resurgence of interest in problems of pattern formation (Connelly et al., 1981) and differential growth processes (Gould, 1977; 1982; Alberch, 1980; 1982) have clearly reestablished the central role for the embryo in evolutionary thinking (Bonner, 1982). While other publications, particularly Butler and Joysey (1978) and Hinchcliffe and Johnson (1980), review data which permit fairly comprehensive evaluations of tooth and limb evolution respectively, here I shall further explicate my previous contention that "Morphological changes in evolution may not be as complex as they first appear. . . ." (Maderson, 1975, p. 316) by exploring how studies of tissue interactions have continued to contribute to this simplification.

ANATOMIC AND DEVELOPMENTAL PATTERNS
IN LIMBS AND DENTAL SYSTEMS

The decision to restrict this discussion to these two systems was predicated on two facts. First, of all vertebrate systems where tissue interactions have been studied, only these two are represented by fossilized remains. Second, the changes in form so documented raise so many problems which can be addressed by developmental biologists that even a discussion of selected problems will occupy valuable space. To facilitate discussion, a proposition of the basic patterns of form is desirable.

The tetrapod limb consists of features arranged in a proximo-distal segmental sequence: dorso-ventral and/or antero-posterior arrangements can also be recognized, but these are ignored here for the sake of brevity. The features comprise the girdle, the stylopod (humerus or femur), the zeuogopod (radius-ulna or tibia-fibula), the wrist/ankle (carpals or tarsals), the palmar/plantar elements (meta-carpals or metatarsals), and the phalanges. The fore/hind foot (manus or pes) comprise the last three named segments collectively referred to as the autopod. The integument may show conspicuous segmental specificity. All muscles run between at least two segments and some may occupy many segments. Nerves and blood vessels run proximo-distally as continuous units, but while their nomenclature may refer sections or branches to particular segments they are not "seg-mented" structures.

In tetrapod dental systems, dermal ossifications laid down around the primordial pterygo-quadrate bars and Meckel's cartilages bear teeth usually arranged in linear series along the maxillae, premaxillae, and the two mandibular rami. "Sets" of the homodont teeth of amphibians and most reptiles are referrable to the dermal bones bearing them, but in mammals we distinguish incisors, canines, and molars. Within various mammalian taxa, each class of tooth varies by number and cusp pattern (in the case of molariform teeth). Again, the integument may show conspicuous regional speci-ficity. Muscles, situated mainly around the suspensorial region, do not occupy such obviously regional locations as do those in the limbs, nor do blood vessels, but the innervation does have some apparent regionality.

The variety of evolutionary and developmental problems posed by these two systems can be simply illustrated. A sprawled modern crocodile has limbs with many skeletal elements (especially in the autopod) and an equally numerous series of individual ossifications and sclerifications in the dental system, the whole covered with a scaled integument, the individual units of which vary only in size and, to some degree, shape. Now consider a stork and a kangaroo. In both comparisons the disparity in limb size relative to the body, and between limb segments, is obvious. Simple counting of the number of ossifications in both limbs and dental systems reveals reduction, greater in the mammal than in the bird. The integument-ary coverings, relatively uniform in the crocodile, present the most striking differences, both within and between the bird and the mam-mal. The stork's wing is covered with feathers, but these have dif-ferent forms over different wing segments, and many, especially the flight feathers, are numerically and topographically defined for that species. Claws are absent from the distal ends of the phalanges. However, on the leg a quite different situation obtains. Plumage

feathers are restricted to the girdles, femur, and upper tibia, while
the lower tibia and distad segments possess scales of diverse form
(Sawyer, this volume). There are terminal claws. The stork's dental
system lacks teeth, but the buccal cavity has distinct transverse
ridges. The bill is conspicuous, but caudal to it, the integument is
pterylous. In the kangaroo equally striking regional specificity of
the integument is apparent. In the adult teeth heterdonty is obvious,
but notice the diastema which, like the freely-flexible adjacent
glabrous lips, is seen in so many eutherian herbivores. Vibrissae,
sweat-glands, sebaceous glands, each with their own domains and
numerical specificity, add to the picture of spatial complexity.

Documentation of the varied expression of form in these two
systems can be found in thousands of texts, but such is not necessary;
any developmental biologist is familiar with the fundamental problem.
He asks, "How does a zygote's gentotype find expression in the adult
phenotype?" This fundamental formulation differs from that concerning the evolutionary biologist only in the time scale involved: the
developmental biologist is concerned primarily with the life cycle of
a single organism beginning with fertilization, ending with death.
The evolutionary biologist sees such life cycles as lying at right
angles to a time scale which, in the case of tetrapod history, began
in the mid-Devonian, approximately 370 million years ago, and has
produced dozens of lineages with distinctive forms by a process
called evolution which is probably still going on and, man permitting,
will probably continue until our solar system is recycled within the
cosmos. The only limitation of this prediction is that we have absolutely no way of estimating what forms might appear in the near and
distant future since we cannot predict the nature of future environ-
ments.

While many nineteenth century zoologists penned some extraordinarily prescient statements (see, for example, T. H. Huxley's
comments on the cell membrane in Moscona, 1974, p. ix), it is
safe to say that were the authors here today, even the most brilliant
would be dumbfounded by the extraordinary range of disciplines which
today constitute zoology: I am sure that many of us share the same
sentiment. Within a specific biological discipline, reductionist
models can ameliorate the sense of intimidation which we all feel
in the face of the avalanche of new data which threatens to bury us.
These reductionist principles in turn facilitate interdisciplinary
communication. I find that the necessary simplification of the problems attendant upon the evolution of complex systems has been enhanced by the proposition of models which attempt to describe their
developmental basis. I wish to state explicitly that my phrasing of
the ensuing discussion of limb and dental system evolution in terms
of Wolpert's Positional Information Hypothesis (1980; 1981; 1982)

does not imply that I believe that this model explains all of development. Ultimately, our understanding of both development and evolution will necessitate an appreciation of the molecular mechanisms involved. Perhaps by the end of this century we will be in a position to talk intelligently in such terms and the potential power of "source-sink diffusion models" (see various papers in Bonner, 1982) will be realized. However, given the current controversies concerning genomic organization and operation (Dover and Flavell, 1982), my time frame might be overly optimistic. This suggests that a conservative approach using Positional Information (which is merely a refinement of classical gradient theory) is more appropriate to the hard data at hand.

The history of our current understanding of limb morphogenesis and the central role that this feature has played in discussion of Positional Information, as well as alternative hypotheses (e.g., the polar coordinate model) are too well-known to warrant explication here. The interested reader will find pertinent references in Fallon et al. (this volume), Wolpert (1980; 1981; 1982) and Bryant et al. (1981). For the present, the proposition that a distally located "progress zone" leaves behind mesenchymal cells which comprise "nonequivalent," segmentally-arranged populations, as limb bud outgrowth proceeds is the most important. Wolpert (1981, p. 14) has related the thesis to the classic explication of neural induction (Toivonen et al., 1976). Our current knowledge of the morphology, let alone the dynamics, of craniofacial development is as yet so inadequate by comparison with that available for limb (Pratt and Christiansen 1981) that it would be premature to attempt to even describe the entire system in terms of any comprehensive schema, let alone propose specific analogies for the progress zone of the limb. However, insofar as the ectomesenchyme (neural crest) is known to play a vital role in craniofacial morphogenesis (Pratt and Christiansen, 1981), it is of interest to note that evidence for "prepatterning" of these cells has been available for over 35 years. Discussing experiments involving rotations of the neural plate of Ambystoma mexicanum larvae, Horstadius (1950, p. 51) concludes: ". . . there is a qualitative difference between the ectomesenchyme of the mandibular arch and that of the gill-arches." Whether this difference is produced by morphogenetic interactions associated with neural induction, or whether it is an emergent property dependent on the microenvironment(s) through which the cells migrate (Noden, 1980a) is relatively unimportant in the present context, it is the existence of patterns per se which is important. Over the past 15 years Noden and others (see references in Noden, 1980a) have described highly specific migratory pathways for ectomesenchymal cells. Replacing premigratory presumptive 2nd branchial arch crest

in chick embryos by midbrain crest from the quail produced supernumerary 1st arch skeletal elements in the 2nd arch location in the
host (Noden, personal communication), a result very similar to
that reported by Horstadius (1950). As Kollar (1982) indicates, the
sequences of tissue interactions involved in tooth-jaw (only parts
of the total complex dental system) morphogenesis are rather complex. Reporting experimental contributions to the problem of molar
patterning in the mouse Lumsden (1979, p. 77) writes: "The results
are consistent with developmental theories which propose that grada-
tions of shape and size in the individual sequentially initiated elements of a series are expressions of intrinsic time-dependent alterations in the growing cell population which forms them." Kollar and
Lumsden (1979) report the role that trigeminal innervation plays in
specifying tooth locus formation along the continuous dental lamina.
Among other points it might be indicated that, since such specification is tantamount to "instructing" specific populations of ectomesenchymal cells to begin differentiating as odontoblasts, these
data suggest a situation par excellence where ectomesenchymal cells
rely on their final microenvironment to become "completely determined." I cannot leave this point without taking the opportunity to
bear responsibility for committing to print a marvelous speculation
which Dr. Kollar made to me recently. Given Noden's (1980b;
1980c) confirmation that the trigeminal ganglion forms from cells of
ectomesenchymal and ectodermal placodal origin, could these latter,
having been "positionally informed" as a spinoff of axial (neural)
determination, ultimately play their role as pattern producers by
becoming the determining neurons? As Charlie Brown would say,
"The theological implications alone stagger the imagination."

In summary, currently available data suggest that limbs and
dental systems develop form as an expression of fundamental patterns. In principle, the multitude of specific forms which their constituent features have manifested throughout evolution can be explained as originating from modulations of the morphogenetic processes which are the causal agents of pattern expression. These
morphogenetic processes are relatively few in number. Differential
mitotic activity between nonequivalent segments obviously influences
the mesenchymal cell aggregates which represent the primordia of
skeletal and muscular features. These aggregates vary in size and
this must be the basis for differential relative size of homologous
features in different species: the factors responsible for specific
shape are as yet unknown (Hall, 1978; 1981). If the number of aggregates laid down in a segment were to change, the number of mature
features would change. Our knowledge of the factors involved in
"individuating" mesenchymal cell aggregates representing chondrogenic anlagen (prerequisites for the eventual presence of individual

skeletal features) has been enhanced by the studies of Melnick et al. (1981) and Jaskoll et al. (1981). These papers document the spatio-temporal patterns of fibronectin distribution in prechondrogenic anlagen in chick limbs and branchial arches, respectively features of endo- and ectomesenchymal origin. Fibronectin was detected in nonrandom "enclosing" patterns around individual primordia prior to the onset of secretion of Alcian blue-positive cartilage matrical materials. These data provide an additional example of a ubiquitous extracellular protein which plays a role in morphogenesis, and a basis for testable hypotheses concerning the way in which modulations of fundamental morphogenetic processes could have produced changes in form of skeletal features during evolution.

Within the context of Wolpert's Positional Information Hypothesis, one aspect of the nonequivalence of mesenchymal cell populations deriving from sequentially produced progress zones is the inductive capacities of the presumptive dermal cells which are responsible for integumentary regionalization (Sengel, 1976). The variation in size and shape of reptilian scales appears to result from dermal influences during development (Dhouailly, 1977). For avian feathers and various types of scales, mammalian hair and nails, and different forms of teeth, there is a body of data which indicates that either dermal papillae or odontoblasts control the epidermis, or a degree of autonomy exists in the latter (Kollar, this volume; Sawyer, this volume; Sengel, 1976). These are evolved inductive systems generating integumentary regionality. Since various segments of limbs and dental systems in mammals and birds show such regionality it follows that different, indeed new, inductive influences are in operation, and these have resulted from uncoupling of developmental events along the proximo-distal axes during evolution from reptilian ancestors.

SELECTED PROBLEMS IN CHANGES IN FORM

Introduction

The forms of limbs and dental systems in known species reveal not only extreme plasticity in pattern expression but also an extraordinary number of examples of convergent form in unrelated lineages. Obvious examples, whose biological roles can be assumed by analogy with living species, are paddle-limbs for swimming or cursorial limbs for speed. Chopping and cutting anterior teeth and a diastema with grinding teeth behind occur in both herbivorous archosaurs and mammals. Carnivorous carnosaurian archosaurs have caniniform teeth and there have existed both metatherian and eutherian saber-tooth tigers. Any attempt to understand how similar environmental pressures can so easily elicit such frequent changes in form should surely seek the answer in modulation of the morpho-genetic factors underlying pattern expression.

Numbers of Skeletal Features in Complex Systems

It has long been realized that from rhipidistians through reptiles, to birds and especially mammals, the number of skeletal features in limbs and dental systems has decreased. In cursorial limbs, loss of digits I and V seems to be associated with a concentration of axial stability through digits III and IV (e.g., artiodactyls) or III (e.g., perissodactyls). The steady reduction in postdentary elements in theriodont reptiles to the single dentary in the mammalian mandible is said to be associated with increased strength, although when relative size is considered, one wonders how a hyena's ability to crunch through a buffalo's femur differs from a parrot's ability to crack open a large seed. However, biological roles aside, we ask how these documented changes in form occurred?

Fossilizable form probably does not accurately represent actual embryonic form of extinct species, any more than do all mature features in a modern species, e.g., muscles and tendons (Lance-Jones, 1979). Paleontologists always allow (perhaps too freely) for nonpreserved cartilage in their restorations. Indeed, cartilaginous features such as the hyoid of most mammals, which do not show replacement ossification (a simple heterochronic modu-lation of a ubiquitous morphogenetic sequence), may have extremely important biological roles and yet be unrepresented in fossils. Many transient mesenchymal aggregations have been described in embryos and have long been the subject of debate in the context of "the prob-lem of vestigiality" (for a recent view and references, see Regal, 1977). Since such debates have not yet satisfied everyone, we should consider whether the issue is amenable to experimental investigation. Perhaps these "vestigial" features facilitate the morphogenesis of other features which do have identifiable biological roles in the adult organism, and here I explicitly extend Bock and von Wahlert's (1965) concepts to embryonic features, offering an example of the utility of this approach.

Data have been cited concerning the spatio-temporal distribu-tion of fibronectin around prechondrogenic primordia. This protein characterizes all skeletal tissues including membrane bones (Jaskoll et al., 1981, p. 211) and its distribution around all skeletal features in the mouse autopod (Jaskoll, personal communication) suggests that it is the morphogenetic agent producing individuation. If the opposing faces of closely juxtaposed prechondrogenic primordia (e.g., in the carpus or tarsus) were "defective" in their fibronectin coat prior to chondrogenesis, fusion would occur: such fusions, and the resulting adult forms of features are well documented (Alberch and Alberch, 1981). On the other hand, if an aggregate had no fibronectin coat, it would not chondrify and would not be identifiable as an adult feature. However, we should not assume that such transient features are "vestigial" in the pejorative sense of the word. They might control the spread of morphogens such as those putatively involved in antero-posterior axial establishment (see references in Hinchcliffe and Johnson, 1980). From this viewpoint, a transient feature could have both form and function and its resultant faculties would have biological roles seen by selection through its contribution to the morphogenesis of feature(s) which a functional anatomist could analyze or a paleontologist describe. Such postulates are amenable to investigation—further analysis of normal differentiation in Bolitoglossa occidentalis (Alberch and Alberch, 1981), comparison of normal chick limb to that of talpid[3] (see references in Hinchcliffe and Johnson, 1980) or in experimentally-induced fused hyperphalangic and/or syndactylous avian feet like those reported by Pautou (1976).

In principle it would appear that modulation of fibronectin activity could just as easily produce an increase in number of skeletal elements, but this is a rare phenomenon. Nowhere is this better illustrated than in paddle-limbs. Contrary to what one might expect from external form (e.g., the graceful, sinuous paddles of many sea-turtles), only three lineages—ichthyosaurs, sauropterygians, and cetaceans—show hyperphalangy and/or polydactyly. Given the "ease" with which modulation of AER activity could have produced webbing (Fallon et al., this volume), an obvious initial character state in a sequence of autopodal forms in a transformation series leading to a paddle, the apparent restriction of increase in number of skeletal features seems puzzling. Until now, that numbers of skeletal features showed a reduction throughout osteichthyan-tetrapod evolution was but a phenomenological fact for which no explanation was available. Let us assume that one aspect of the nonequivalence of populations of mesenchymal cells produced by progress zones within a given system is canalization to produce a certain number of aggregates which can be easily reduced, but only increased with great difficulty since they require major changes in a polygenic system. This postulate finds circumstantial support in the development and evolution of the avian wing (Maderson et al., 1982, pp. 17-19) but more direct evidence is available. The increased number of chondrification centers in talpid[3] are only one of numerous pleiotropic effects of this lethal allele and seem to result from changes in membrane function. Such functions are obviously crucial to all aspects of endomesenchymal differentiation perhaps telling us that such genetic changes exemplify the "absolute developmental constraints" discussed by Maderson et al. (1982, p. 288). This explanation is supported by Lande's (1980, p. 234) discussion of the relative probabilities of regulatory versus structural gene mutations appearing and/or becoming fixed in natural populations. The alleged "macromutations" having significant biological roles (Maderson, 1975, pp. 324-325) do not have to be such: the scenario depicted as producing "ultimate perfection" in morphological features could be produced by exactly the type of genetic changes which Lande (1980) discusses. Once more, analysis of fibronectin activity in mutants and phenocopies could elucidate the cellular mechanisms concerned, but pertinent data are already available. In many experiments whole or partial duplications of sets of limb features have been described. I am informed (Saunders, personal communication) that in such experiments totally unidentifiable features are rarely if ever seen. These phenocopies seem to mimic the mutant eudiplopod (Hinchcliffe and Johnson, 1980, p. 119) which, while it has low penetrance and thus produces extremely variable form of the wing, has clearly identifiable individual features. All of these facts pertain to "canalization," a phenomenon which will now be discussed with particular reference to the "adaptive radiation," or lack thereof, of the avian wing.

Heterochrony and Canalization in Avian Wing Evolution

Any of the extant 9,000 avian species can be recognized as a bird because of the overall adaptation of the body to the biological role of controlled flight (Feduccia, 1980). There is a curious fact about the skeletal features of flightless species. Only moas lack all features beyond the girdles, only the kiwi shows extensive reduction and/or simplification, the rest differ from flying birds only in the relatively reduced wing size which Feduccia (1980) relates to neoteny. Why is it apparently so difficult for a bird to lose its wings? Why is there no loss of distal musculoskeletal features as is so common in lizards (Gans, 1975)?

If all reduced wings have significant biological roles such as the distinctive flapping of ostriches, pertinent documentation is lacking. If they are so small that selection no longer sees them so that they are neutral features whose persistence is an historical accident, we are surely seeing a highly improbable number and degree of parallelisms. Avian wings have several unique developmental characteristics (Fallon et al., this volume). I propose that during protoavian evolution, selection for aerodynamic efficiency in features contributing to controlled flight was so rigorous that variability in form was limited to such an extent that we now see an extraordinarily highly canalized developmental system. The wing is thus a microcosmic "Bauplan" offering a unique model to explore the genetic basis of an otherwise phenomenological concept (Maderson et al., 1982). Quantitative alterations of the polygenic system underlying morphogenesis are expressed in heterochronic growth producing relatively reduced size of the feature. Integumentary development can be, and has been, easily decoupled producing changes in the form of flight feathers. Qualitative changes in form and function of musculoskeletal features cannot occur. Lande (1980) notes that deleterious pleiotropisms usually accompany major allelic changes in polygenic systems and such occur during artificial selection for wing reduction. Of the three Mendelian wingless strains known, two are sex-linked (Pease, 1962) and the third not only has a very low penetrance but has many pleiotropisms which would be lethal in nature (Waters and Bywaters, 1943). Fallon's "limbless chick" (personal communication) has the mutant allele producing only a defective AER but, apart from the animals' need to be removed from the egg, such a "hopeful monster" would have a very poor chance of mating.

Diastemal Evolution: the Limits of Paleontology

Even before Permo-Triassic theriodonts had evolved true heterodonty some, and even contemporary "quantitatively heterodont" pelycosaurs, developed diastemae. Late Triassic mammalian tricondodonts (which have been variously suggested to have been insectivorous or omnivorous) and the Mesozoic "rabbits" (herbivorous multituberculates) expressed this form. However, it is in the mid-Cenozoic ungulate lineages that diastemae, with or without hyper- or hypotrophic incisors and/or canines, appear with astounding frequency. Considering turtles, several archsaurs, 99.9 percent of known birds, and sundry mammalian lineages, we see that loss of some or all the teeth is almost a tetrapod characteristic.

The work of Kollar and others (see Kollar, this volume) suggests that whole or partial tooth loss could have been achieved by modulation of a variety of morphogenetic mechanisms: changed or lost competence of migrant ectomesenchymal cells, of adjacent dermal bones, of superficial ectoderm influencing the development of dermal bones, changes in innervation (as discussed above), or loss of competence of buccal ectoderm. We can never know which changes occurred in extinct lineages, but there is a sufficiency of extant models to explore many of these possibilities.

One specific problem deserves comment. While supernumerary teeth within a class are occasionally seen, missing teeth are very rare. How then can we account for the apparent suddenness of diastemal appearance in so many Cenozoic ungulates? The fossil evidence seems to demand macromutations which were successful in lineages adapting to herbivorous feeding. Anatomical studies on perinatal rodents and lagomorphs (Moss–Salentijn, 1978) show that vestigial teeth occur frequently and thus it is unnecessary to assume that tooth loss must be sudden. Mammologists should reexamine their specimens to ascertain when evidence of sockets' diminution and disappearance occurred, while study of the tissue interactions leading to socket formation, especially in mutant rodents (Miller, 1978), demands attention.

Integumentary Evolution: Decoupling within Patterned Systems and the Evolutionary Origin of Specific Morphogenic Interactions

The integument of reptilian limbs and dental systems is grossly uniform: where appendages exist on scales we have little direct knowledge of factors controlling their differentiation (Maderson, 1971; 1972a; 1982). In birds and mammals, however, specialized peidermal features occur, some on one or more segments, others on only certain parts of certain segments. Reptilian and avian scales arise primarily due to autonomous epidermal capacities (Dhouailly, 1977; Sawyer, this volume), but hairs, feathers, nails, glands, etc. whose protein synthetic expressions may represent mere quantitative modulations of fundamental and general epidermal capacities (Flaxman, 1972; Flaxman and Maderson, 1976) are examples of specialized features arising from changed morphogenetic sequences. Models for the evolution of hair (Maderson, 1972a) and feathers (Maderson, 1972b; Regal, 1975; Downie, 1982), dealing only with the form of individual features, their patterned topographic distribution not having been addressed except indirectly in the context of the premise of their origin from scaled integuments (Maderson, 1972a; 1972b), note the remarkable evolution of convergent form in the presence of subjacent dermal papillae. These expressions of embryonic form (which of course in these examples persist into adult life) exemplify the need for rigor in the use of terms to describe the "actions and/or activities" of populations of cells if one seeks a precise evaluation of both morphogenetic and evolutionary problems.

For feather morphogenesis Sengel (1976) reviews the sequence of cytological events wherein presumptive dermal mesenchyme first becomes densified, and subsequently localized condensations form the primordia of the dermal papillae. Sengel (1976) and Sawyer (this volume) document the facts that: (1) such condensations are absent from apterous regions; (2) feathers can be induced in normally apterous regions by physical manipulation of the early somatopleur; (3) condensations are absent from most of the body of scaleless mutants, but where present induce randomly arranged feathers; and (4) such condensations are absent from the anlagen of all avian scales. Dhouailly et al. (1980) have reported that feathers can be induced to differentiate at the tips of scutate and scutellate scales by the in ovo application of retinoic acid: the result is time-dependent and the best results are obtained when the drug is applied when there exists transient mesenchymal condensations beneath the distal tip of the epidermal placode (Dwyer, 1971). These and other data permit speculation on the evolutionary origin of specific tissue-interacting systems.

Studies of epithelial-mesenchymal interactions have featured limb and skin as their most popular models for over 40 years. Students of the limb have been spared the semantic horrors which have bedevilled skin studies for over a century. Feathers and hairs have obvious dermal papillae, and teeth have equally clear "embryonic form." During the flowering of Haeckelian investigation, Kerbert (1876) studied various developing features of the amniote integument and concluded that "hairs, feathers and scales exactly resemble one another in their early development." By the time Billingham and Silvers wrote their seminal review in 1963, everything growing from the surface of the body was an appendage and subsequently, via extrapolation from otherwise good experimental data for hairs, feathers, and teeth, the dermal mesenchyma was thought to control everything. Anyone interested in understanding the chaos and confusion deriving from Kerbert's misguided preconceptions should read Dr. Sawyer's (this volume) references in chronological order.

That dermal papillae and dental mesenchyme give highly specific instructions to the overlying ectoderm in the morphogenesis and homeostasis of hairs, feathers, and teeth is indisputable (Dhouailly, 1977; Kollar, this volume; Sawyer, this volume). That lizard mesenchyme controls scale shape is also indisputable (Dhouailly, 1977). However, in scale development and also in the development of palmar/plantar integument in man where, it is alleged, the dermis controls differentiation and adult homeostasis (see discussion, Sengel, 1976), the lack of recognizable embryonic form is strikingly different from papillae. Thus, in the absence of many critical data, we should be cautious in generalizing the mechanisms underlying the genesis of all epidermal features.

Within the conceptual framework of Positional Information, the evolution of changed form of epidermal features must have in-

volved specification of superficial mesenchyme to engage in different
types of interaction with the epidermis overlying each segment of
the limb and dental system. I will confine my comments to hairs,
feathers, and scales, leaving analysis of teeth to Kollar (this volume)
and integumentary glands to Cuhna et al. (this volume) and Hardy
(this volume).

What gives "mesenchymal condensates" their extraordinary
faculties? Kollar and Kerley (1980) have reported that, once determined, odontoblasts retain their peculiar faculties after even the
most strenuous experimental attacks. Oliver (personal communication) has found that in culture, vibrissal papillar cells continue to
behave differently from other connective tissue cells being "reluctant
dividers and spreaders and seem to be happiest when heaping up on
top of one another." Fisher (personal communication) finds that when
"perivibrissal mesenchyme" is cultured as an organ explant, the
behavior of the cells differs depending on the method of separation
used to remove the ectoderm: after EDTA separation (which leaves
most of the basement membrane intact), the vibrissal papillae
retain their integrity, but after trypsin separation (which destroys
most of the basement membrane), the papillar cells clump together
to form "super-papillae." Available data on avian feather versus
scutate/scutellate scale development, when judged from embryonic
form alone, suggest that feathers "need" to be induced by mesenchymal condensates which are normally absent from scale primordia.
One wonders whether study of the morphogenesis of retinoic acid-
induced feathers (Dhouailly et al., 1980) would reveal a prolongation
of the existence of Dwyer's (1971) "transient" condensations. This is
predictable since other studies have shown that retinoic acid affects
matrical molecular metabolism (Sauer and Evans, 1980), which
molecules are associated with mesenchymal cell aggregation (Pratt
and Christiansen, 1981). This experimental material should be compared to "feathered-scale" morphogenesis which is a normal feature
of many domestic breeds. Such studies could contribute to the models of feather evolution from scales which have been proposed and
lay a foundation for future evaluation of how the ubiquitous matrical
molecules might have been involved in evolutionary change by modulating cellular behavior.

## "CHICKENS' TEETH" AND PLANS FOR THE FUTURE

Having reached the point where we can plan experiments using
the techniques of developmental biology to address evolutionary
questions directly (Maderson et al., 1982), we should be aware that
some traditional dogmas will be challenged and new conceptual

problems will need to be addressed. Dhouailly's (1977) many elegant experiments involving xenoplastic recombinations of dermis and epidermis among amniotes dramatically reinforce the classic contention that tissues can only respond within the limits of their genetic competence. Kollar and Fisher's (1980) "chickens' teeth" seem to be exceptions to this general rule and warrant detailed discussion.

In light of the range and importance of the biological issues involved I am personally too impatient to question whether Kollar and Fisher produced teeth: the burden of proof should be borne by those who have implicitly or explicitly expressed doubts. Indubitably avian ectoderm interacted with competent mesenchyme to do something far more complex than form simple buccal epithelium. Previous caveats can be abandoned and the answer to a previous rhetorical question: "Does evolution . . . involve irrevocable changes in genetic competence?" (Maderson, 1975, p. 317) must here receive a negative response. This must inevitably disturb population geneticists: the appropriate alleles should have been lost by genetic drift in a relatively short period of geologic time especially in birds which do not have conspicuously large breeding populations. Drift is of course only a probabilistic phenomenon, but the fact that "vestigial" dental laminae have been described in several different avian lineages (Romanoff, 1960, pp. 459-460) implies that several different violations of probability have occurred. A molecular biologist could argue that the alleles might have been conserved because their transcripts produced proteins used other than in tooth differentiation. This point has validity in light of the demonstration of the apparent evolutionary conservatism of amelogenic proteins as demonstrated by immunologic comparisons (Herold et al., 1980). However, this explanation ignores important issues: we may not arbitrarily restrict identification of form to a molecular level since selection sees molecules only indirectly via their contributions to features which result from morphogenesis. Having elected to work within a conceptual framework based on a hierarchy of units—genes, proteins, cells, tissues, organs, and organisms (Salthe, 1975), with an associated hierarchy of causal mechanisms linking all levels, neither a developmental nor an evolutionary biologist can ignore the fact that a very complex sequence of events produced these chickens' teeth.

Previous studies (Kollar, this volume) on mammalian material demonstrate that it is the odontoblastic mesenchyme which specifies tooth shape. Knowing that outside the extinct theriodont reptiles homodonty is the rule and the possible explanations for the evolution of the genome of the avian ectoderm discussed above, two equally plausible predictions could have been made concerning the form of the teeth which Kollar and Fisher (1980) set out to grow. If indeed the entire polygenic system underlying the primitive avian ectoderm's involvement in tooth morphogenesis had been retained in a quiescent state, no matter what the origin of the xenoplastic dental mesenchyme, only homodont teeth should have been formed. Alternatively, if only those genes potentially capable of permitting the production of amelo- genic proteins had been retained, the ectoderm should have behaved like any competent mammalian ectoderm and participated in the formation of molar teeth in the presence of the appropriate inductive stimulus. Neither prediction was fulfilled; some experimental teeth were molariform, others were homodont.

Since there is not a shred of evidence that any archosaur ever had molariform teeth, an evolutionary anatomist would deny the possibility of chickens' teeth expressing such a form. Thus, Kollar and Fisher (1980) caused two major "rules" to be broken. First, in clear violation of the dogma of genetic drift, a xenoplastic inductive message caused avian ectoderm to pursue a developmental pathway which it has not pursued for about 100 million years—cytogenesis manifested in amelocyte differentiation. Second, in violation of a basic rule of tissue interactions, the ectoderm pursued a develop- mental pathway which it had never expressed in its entire evolution- ary history when it participated in molariform organogenesis. Some months ago an undergraduate asked me if there were ways to make a snake or legless lizards' limb bud continue to develop and if so, what would one expect to obtain? Idly turning over in my mind many of the data which other contributors to this book have published, and garnishing this delectable potpourri with a soupçon of Positional Information, I realized that Kollar and Fisher's (1980) chickens' teeth were but one manifestation of a fascinating problem which will confront practitioners of paleodevelopmental biology. Inasmuch as all tissue manipulation data reveal interactions, and many believe that Positional Information is everywhere, I formulated the following thought: perhaps Kollar and Fisher's data imply that when an entire complex anatomical feature is reconstructed, as distinct from the parts which are customarily examined in tissue recombinant studies, maybe the rules change. If this were so then the practice of experi- mental paleodevelopmental biology could lead us not only to an under- standing of how developmental events become uncoupled during the evolution of complex features, but also could yield insights into the causality of nonequivalence (Wolpert, 1981).

Only the single paper of Jorqueora and Pugin (1971) reports positive success from interclass recombinations of whole limb com- ponents. Their reciprocal exchanges of limb ectoderm and mesoderm from rat and chick embryos produced some "normal" chimaeric limbs in the sense that customary developmental rules were obeyed, but others were implied to have some changes in both ectodermal and mesenchymal derivatives. Saunders (personal communication) tells me that in his hands neither such recombinations nor similar turtle-chick recombinations develop, while Kollar and Fisher (1980) report that mouse-quail recombinations failed to produce teeth. The data concerning differential behavior of vibrissal dermal papillar cells following different technical manipulations, as discussed above, suggest that interclass recombinations should be feasible. Certainly the endeavor would be worthwhile.

Reports on intraclass recombinations are more numerous than interclass data, but the results are tantalizing for other reasons. The classic example of restricted genetic competence to respond to a heteroplastic inducer is Rotman's (1939) study of ectopic optic vesicle transplants between two species of the urodele Triturus: the induced lens' shape and size corresponded to that of the host, not the inducing graft. However, subsequent investigations of other amphibian species (see references in Reyer, 1977) revealed that in certain cases considerable accommodation could be obtained. It is unclear whether these data reflect the peculiarities of timing of neural induction processes per se, or whether all experiments should be repeated with the sort of attention given to embryonic age or even environmental temperature that is routine in chick work.

A single experiment documenting results exactly resembling those reported by Kollar and Fisher (1980) comes from Hiyashi (discussed by Saunders, 1970, p. 120). Heterotypic duck and chick ectoderm will form "duck-type tooth [sic] ridges" when combined with duck beak mesenchyme. Chick leg ectoderm placed in contact with duck leg mesoderm produces a webbed form (Sengel, 1976) although it is probably incorrect to claim that an exact replica of normal duck webbing occurs (Saunders, personal communication). Heterospecific recombinations of chick and duck tissues produce chimaeric feathers: "whose barbules are in conformity with the specific origin of the epidermis, but whose general architecture (number of barbs, presence or absence of rachis, shape, size) is in conformity with that of the dermis" (Dhouailly, 1977, p. 106) (emphasis added). I am at a loss to decide how such details should be evaluated vis-a-vis Kollar and Fisher's data except to comment that they exemplify the problems in definition of categories of form which will need to be addressed in the future whether the questions being asked are developmental or evolutionary. While reciprocal transplants between avian leg and wing tissues (Cairns and Saunders, 1954) definitely produce extensive modulation of normal expression (e.g., leg ectoderm on wing mesoblast does not form scales) none has documented the precise form of the integument with respect to number and spatial distribution of the flight feathers. Indeed one might question whether the wing is an appropriate model to examine this degree of refinement of inductive potential since it obviously

possesses some very unique features (Fallon et al., this volume).
One wonders how both mesenchymal and ectodermal cell populations
would respond in fore and hind limb reciprocal embryonic exchanges
in a lagomorph, canid carnivore, or caviid roden where the number
of skeletal features in the autopod, and its superficial integuments
(e.g., claws, dew-claws, and pads) differ.

Xenoplastic grafting of blocks of mesenchyme under host
mesoderm (Cairns, 1965), or recombined heteroplastic (Dhouailly,
1977) or homoplastic and/or heterochronic (Sengel, 1976) grafts
grown on chorioallantoic membranes (CAMs) certainly indicate the
role of mesenchyme instructing epithelial differentiation, and the
limits thereof, either as a result of "genetic restriction" or age-
determined competence. In reciprocal heterotopic recombinations
there is no doubt that the form of individual features such as hairs,
feathers, or avian scales of various types can be recognized. However, I have always been struck by the relatively irregular groupings
of such experimentally induced features. In the future, more attention
should be paid to this aspect of mesenchymal instruction.

Sawyer's (this volume) review of the work on the scaleless
mutant indicates the wide range of anatomical, histological, ultrastructural, and biochemical data which are needed to analyze the
genesis of form for a single feature: the tissue recombination techniques he discusses give us an insight into the morphogenetic mechanisms (faculties of embryonic form) which underlie their expression,
while the genetic studies afford glimpses of the most puzzling aspects
of both morphogenesis and phylogenesis. Inasmuch as we postulate
that selection sees the entire organism in, for example, a bird,
where it is intuitively obvious that aerodynamic efficiency is a significant biological role contributing to fitness, we must consider all
aspects of the form of the wing. A developmental biologist, studying
the details of emergent form (e.g., the number, distribution, and
exact location of flight feathers along the postaxial margin of the
zeugopod and autopod) can provide, with no special additional effort,
data which can help an evolutionary biologist understand the evolutionary process. Those of us who are interested in tissue interactions
are in a unique position to provide data which are of interest in a
variety of biological disciplines.


ACKNOWLEDGEMENTS

I wish to thank the many colleagues who have offered information
and advice in the preparation of this work. The responsibility for
speculations is mine. Ms. Judith Steinberg typed the manuscript.
Research support comes from NIH NS-13924 and RR-07119.

REFERENCES

Alberch, P. (1980). Ontogenesis and morphological diversification. Am. Zool. 20:653-667.

Alberch, P. (1982). Developmental constraints in evolutionary processes. In Evolution and Development (J. T. Bonner, ed.), pp. 313-332. Springer-Verlag, Heidelberg.

Alberch, P. and Alberch, J. (1981). Heterochronic mechanisms of morphological diversification and evolutionary change in the neotropical salamander. Bolitoglossa occidentalis (Amphibia, Plethodontidae). J. Morphol. 167:249 (Abstract).

Billingham, R. E. and Silvers, W. K. (1963). The origin and conservation of epidermal specificities. N. Engl. J. Med. 268:477-480 and 539-545.

Bock, W. J. and von Wahlert, G. (1965). Adaptation and the form-function complex. Evol. 19:269-299.

Bonner, J. T. (ed.) (1982). Evolution and Development. Springer-Verlag, Heidelberg.

Bryant, S. V., French, V., and Bryant, P. J. (1981). Distal regeneration and symmetry. Science 212:993-1002.

Butler, P. M. and Joysey, K. A. (eds.) (1978). Development, Function and Evolution of Teeth. Academic Press, New York.

Cairns, J. M. (1965). Development of grafts from mouse embryos to the wing bud of the chick embryo. Dev. Biol. 12:36-52.

Cairns, J. M. and Saunders, J. W. (1954). The influence of embryonic mesoderm on the regional specificity of epidermal derivatives in the chick. J. Exp. Zool. 127:221-248.

Cameron, J. A. and Fallon, J. F. (1977). The absence of cell death during development of free digits in amphibians. Dev. Biol. 55:331-338.

Connelly, T. G., Brinkley, L. L., and Carlson, B. M. (eds.) (1981). Morphogenesis and Pattern Formation. Raven Press, New York.

Dhouailly, D. (1977). Dermo-epidermal interactions during morphogenesis of cutaneous appendages in amniotes. Front. Matr. Biol. 4:86-121.

Dhouailly, D., Hardy, M. H., and Sengel, P. (1980). Formation of feathers on chick foot scales: a stage-dependent morphogenetic response to retinoic acid. J. Embryol. Exp. Morphol. 58:63-78.

Dover, G. A. and Flavell, R. B. (eds.) (1982). Genome Evolution and Phenotypic Variation. Academic Press, New York (in press).

Downie, J. R. (1982). The origin of feathers: deductions from development. Q. Rev. Biol. (submitted).

Dwyer, N. (1971). "Chick scale morphogenesis: early events in the formation of overall shank pattern and individual scale shape." Unpublished Masters Thesis, University of Massachusetts, Amherst.

Eldredge, N. and Gould, S. J. (1972). Punctuated equilibria: an alternative to phyletic gradualism. In Models in Paleontology (T. J. M. Schopf, ed.), pp. 82-115. Freeman, Cooper and Co., San Francisco.

Feduccia, A. (1980). The Age of Birds. Harvard University Press, Cambridge.

Flaxman, B. A. (1972). Cell differentiation and its control in the vertebrate epidermis. Am. Zool. 12:13-25.

Flaxman, B. A. and Maderson, P. F. A. (1976). Growth and differentiation of skin. J. Invest. Dermatol. 67:8-14.

Gans, C. (1975). Tetrapod limblessness: evolution and functional corollaries. Am. Zool. 15:455-467.

Gans, C. (1979). Momentarily excessive construction as the basis for protoadaptation. Evol. 33:227-233.

Gould, S. J. (1977). Ontogeny and Phylogeny. Belknap Press, Cambridge.

Gould, S. J. (1982). Change in developmental timing as a mechanism of macroevolution. In Evolution and Development (J. T. Bonner, ed.), pp. 333-346. Springer-Verlag, Heidelberg.

Hall, B. K. (1978). Developmental and Cellular Skeletal Biology. Academic Press, New York.

Hall, B. K. (1981). Intracellular and extracellular control of the differentiation of cartilage and bone. Histochem. J. 13:599-614.

Herold, R. C., Graver, H. T., and Christner, P. (1980). Immuno-histochemical localization of amelogenins in enameloid of lower vertebrate teeth. Science 207:1357-1358.

Hinchcliffe, J. R. and Johnson, D. R. (1980). The Development of the Vertebrate Limb: an Approach through Experiment, Genetics and Evolution. Clarendon Press, Oxford.

Horstadius, S. (1950). The Neural Crest. Oxford University Press, New York.

Jaskoll, T., Melnick, M., MacDougall, M., Brownell, A. G., and Slavkin, H. C. (1981). Spatiotemporal patterns of fibronectin distribution during embryonic development. II. Chick branchial arches. J. Craniofac. Gent. Dev. Biol. 1:203-212.

Jorqueora, B. and Pugin, E. (1971). Sur le comportement du meso-derme et de l'ectoderme du bourgeon de membre dans les echangees entre le Poulet et le Rat. C.R. Acad. Sci. [D] (Paris) 272:1522-1525.

Kerbert, C. (1876). Ueber die Haut der Reptilien und anderer Wirbelthiere. Arch. Mikr. Anat. 13:205-262.

Kollar, E. J. and Fisher, C. (1980). Tooth induction in chick epi-thelium: expression of quiescent genes for enamel synthesis. Science 207:993-995.

Kollar, E. J. and Kerley, M. A. (1980). Odontogenesis: interactions between isolated enamel organ epithelium and dental papilla cells. Int. J. Skelet. Res. 6:142-148.

Kollar, E. J. and Lumsden, A. G. S. (1979). Tooth morphogenesis: the role of innervation during induction and pattern formation. J. Biol. Buccale 7:49-60.

Lance-Jones, C. (1979). The morphogenesis of the thigh of the mouse with special reference to tetrapod muscle homologies. J. Morphol. 162:275-310.

Lande, R. (1980). Microevolution in relation to macroevolution. Review of Macroevolution: Pattern and Process (by S. M. Stanley, W. H. Freeman and Co., San Francisco, 1979), Paleobiology 6:233-238.

Lovtrüp, S. (1981). The epigenetic utilization of the genomic message. In Evolution Today (G. G. E. Scudder and J. L. Reveal, eds.), pp. 145-161. Hunt Institute for Botanical Documentation, Carnegie-Mellon Univ., Pittsburgh.

Lumsden, A. G. S. (1979). Pattern formation in the molar dentition of the mouse. J. Biol. Buccale 7:77-104.

Maderson, P. F. A. (1967). A comment on the evolutionary origin of vertebrate appendages. Am. Nat. 101:71-78.

Maderson, P. F. A. (1971). The regeneration of caudal epidermal specializations in Lygodactylus picturatus keniensis (Gekkonidae, Lacertilia). J. Morphol. 134:467-478.

Maderson, P. F. A. (1972a). When? Why? and How? Some speculations on the evolution of the vertebrate integument. Am. Zool. 12:159-171.

Maderson, P. F. A. (1972b). On how an archosaurian scale might have given rise to a feather. Am. Nat. 106:424-428.

Maderson, P. F. A. (1975). Embryonic tissue interactions as the basis for morphological change in evolution. Am. Zool. 15:315-327.

Maderson, P. F. A. (1982). Some developmental problems of the reptilian integument. In Biology of the Reptilia-Development (F. Billet and C. Gans, eds.), Academic Press, New York (in press).

Maderson, P. F. A., Alberch, P., Goodwin, B. C., Gould, S. J., Hoffman, A., Murray, J. D., Raup, D. M., Dericoles, A., Seilacher, A., Wagner, G. P., and Wake, D. B. (1982). Role of development in macroevolutionary change. In Evolution and Development (J. T. Bonner, ed.), pp. 279-312. Springer-Verlag, Heidelberg.

Mayr, E. and Provine, W. B. (eds.) (1980). The Evolutionary Synthesis: Perspectives on the Unification of Biology. Harvard University Press, Cambridge.

Melnick, M., Jaskoll, T. F., Brownell, A. G., MacDougall, M., Bessem, C., and Slavkin, H. C. (1981). Spatio-temporal patterns of fibronectin distribution during embryonic development. I. Chick limbs. J. Embryol. Exp. Morphol. 63:193-206.

Miller, W. A. (1978). The dentitions of tabby and crinkled mice (an upset in mesodermal-ectodermal interaction). In Development, Function and Evolution of Teeth (P. M. Butler and K. A. Joysey, eds.), pp. 99-110. Academic Press, New York.

Moscona, A. A. (ed.) (1974). The Cell Surface in Development. J. Wiley and Sons, New York.

Moss-Salentijn, L. (1978). Vestigial teeth in the rabbit, rat and mouse; their relationship to the problems of lacteal dentitions. In Development Function and Evolution of Teeth (P. M. Butler and K. A. Joysey, eds.), pp. 13-30. Academic Press, New York.

Noden, D. M. (1980a). The migration and cytodifferentiation of cranial neural crest cells. In Current Research Trends in Prenatal Craniofacial Development (R. M. Pratt and R. L. Christiansen, eds.), pp. 3-25. Elsevier/North-Holland, New York.

Noden, D. M. (1980b). Somatotopic and functional organization of the avian trigeminal ganglion: an HRP analysis in the hatchling chick. J. Comp. Neurol. 190:405-428.

Noden, D. M. (1980c). Somatotopic organization of the embryonic chick trigenemal ganglion. J. Comp. Neurol. 190:429-444.

Pautou, M. P. (1976). Morphogenesis of chick embryo foot as studied by Janus Green B induced malformations. J. Embryol. Exp. Morphol. 35:649-666.

Pease, M. S. (1962). Wingless poultry. J. Hered. 53:109-110.

Pooley, A. C. and Gans, C. (1976). The nile crocodile. Sci. Am. 234:114-124.

Pratt, R. M. and Christiansen, R. L. (eds.) (1981). Current Trends in Prenatal Craniofacial Development. Elsevier/North Holland, New York.

Regal, P. J. (1975). The evolutionary origin of feathers. Q. Rev. Biol. 51:35-66.

Regal, P. J. (1977). Evolutionary loss of useless features: is it molecular noise suppression? Am. Nat. 111:123-133.

Reyer, R. W. (1977). The amphibian eye: development and regeneration. In The Visual System in Vertebrates (F. Crescitelli, ed.), pp. 309-390. Springer-Verlag, New York.

Romanoff, A. L. (1960). The Avian Embryo. Macmillan, New York.

Rotman, E. (1939). Der Anteil von Indukter und regierendem Gewebe an der Entwicklung der Amphibienlinse. Roux Arch. Entw.-mech. 139:1-49.

Salthe, S. N. (1975). Problems of macroevolution (molecular evolution, phenotype definition, and canalization) as seen from a hierarchical viewpoint. Am. Zool. 15:295-314.

Sauer, G. J. R. and Evans, C. A. (1980). Hypervitaminosis A and matrix alterations in maxillary explants from 16 day old rat embryos. Teratology 21:123-130.

Saunders, J. W. (1970). Patterns and Principles of Animal Development. Macmillan, London.

Sengel, P. (1976). Morphogenesis of Skin. Cambridge University Press, Cambridge.

Stanley, S. M. (1979). Macroevolution: Pattern and Process. W. H. Freeman and Co., San Francisco.

Toivonen, S., Tarin, D., and Saxen, L. (1976). The transmission of morphogenetic signals from amphibian mesoderm to ectoderm in primary induction. Differentiation 5:49-55.

Waters, N. F. and Bywaters, J. H. (1943). A lethal embryonic wing mutation in the domestic fowl. J. Hered. 34:213-217.

Wessels, N. K. (1982). A catalogue of processes responsible for metazoan morphogenesis. In Evolution and Development (J. T. Bonner, ed.), pp. 115-154. Springer-Verlag, Heidelberg.

Wolpert, L. (1980). Positional information and pattern formation in limb development. In Current Research Trends in Prenatal Craniofacial Development (R. L. Christiansen and R. M. Pratt, eds.), pp. 89–102. Elsevier/North-Holland, New York.

Wolpert, L. (1981). Positional information, pattern formation and morphogenesis. In Morphogenesis and Pattern Formation (T. J. Connelly, L. L. Brinkley, and B. M. Carlson, eds.), pp. 5–20. Raven Press, New York.

Wolpert, L. (1982). Pattern formation and change. In Evolution and Development (J. T. Bonner, ed.), pp. 169–188. Springer-Verlag, Heidelberg.

# INDEX

Achondroplasias, 192
Adaptive radiation, 227
Adhesion defect, 138
Agglutination, 138
Agglutinin
  $^3$H-acetyl wheat germ, 138
  wheat germ, 138
Alcian blue positive, 179
Ambystoma mexicanum, 222
Amelocyte, 233
Amelogenic proteins, 232, 233
Amniotes, 3, 4, 147
Anchor filaments, 138, 152
Androgen receptor, 52, 55-56, 61
Annular tympanic cartilage,
  195
Antisera to alpha and beta
  keratins, 133-34
Apical ectodermal ridge (AER
  ridge), 4, 5, 6, 7, 11, 13,
  14, 16, 17, 18, 19, 22,
  191, 226
Apodiformes, 22
Apteric
  epidermis, 124, 128
  regions, 230
Apterium, midventral, 154
Apterygiformes, 21
Archosaurs
  carnivorous carnosaurian,
    224
  herbivorous, 224
Artiodactyls, 225
Autofluorescence, 174
Autopod, 220, 225
Avian
  wing, 226, 227,
  scales, 116, 118, 129, 133,
    134, 136, 229

Axis elongation, 6

Backskin, 138
  epidermis, 124, 126-27, 133
Bantam of Pekin, 150, 155
Barb ridges, 124, 148
Barbs, 234
Basal cell carcinomas, 66
Basal lamina, 30, 33, 44, 103,
  105, 167, 169, 170, 171,
  174-75, 177, 179, 204, 205,
  206
  gaps, 183
Basement membrane, 29, 31,
  43, 83, 84, 95, 97, 100,
  105, 106, 137, 175
Bauplan, 228
Beak epidermis, 124
Bill, 221
Binding sites, 138
Biochemical, 121
  anomalies, 137
Biological role, 215, 217, 225,
  226, 227, 228, 235
  simple approximations of, 217
Bipotential, 155
Birds, 21, 22, 147
Bolitogloss, 226
Bone matrix, 77, 84, 86
Bones
  ectopic, 189, 202
  mandibular, 199
  membrane, 189, 200, 201, 225
  skull, 189, 197
Brahmas, 150, 154
5-bromodeoxyuridine, 156

Canalization, 226, 227, 228
Canid carnivore, 235

Carcinogen treated bladders, 66
Cartilage, 32, 79
  annular tympanic, 195
  columella, 189, 195
  ectopic, 189, 196, 202
  external ear, 189
  limb, 189
  Meckel's, 189, 195, 196, 205, 220
  nasal capsular, 189
  tumor, 189, 197
  vertebral, 189, 190
Caviid rodent, 235
Cell, 93
  contacts, 31, 170, 182
  death, 4, 5-6, 218
  epithelial, 167
  glutaraldehyde fixed suspensions, 98
  sorting, 87
  surfaces, 86-87, 93, 138
Cervix, 57
Cetaceans, 226
Chelating agent, 137
Chick, 175, 176, 178
  dermis, 177, 182, 184
  embryo skin, 174
  -mouse recombinations, 178
Chicken, 21, 182
  epidermis, 163
  teeth, 231
Chimaeric
  feathers, 234
  limbs, 233
Chiroptera, 22
Chondrodystrophies, 193
Chondrogenic anlagen, 223
Chondroitin sulfate, 80
Chondronectin, 84
Chorioallantoic membrane, 175, 176, 182
Chorionic epithelium, 121, 126, 134

Claws, 220
  scaleless, 136
Cleft palate, 42
Collagen, 29, 30, 77, 78, 79, 82, 83, 86, 94, 95, 96-97, 98, 100, 104, 105, 106, 137, 177, 193
  defective lattice, 137
  epithelial, 204
  fibers, 175
  gels, 95, 103, 104
  fibrils, 175
  rafts, 94
Collagenase, 79
Colony formation, 100
Columella cartilage, 189, 195
Comparative anatomy, 218
Competence, 136, 155, 171, 172, 232
  epidermal, 156
Complex anatomical units, 217
Conjunctival papillae, 118
Cornea, 32, 106
Corneal epithelium, 124
Cranio-facial development, 222
Crocodile, 218, 220
Cutaneous appendages, 147
Cytoskeleton, 29
Cytodifferentiation, 35, 51, 63, 68

Definitive scale ridge, 136
Dental
  arch, 28
  defects, 41
  development, 44
  lamina, 42
  mesenchyme, 230
  system, 216
Dentary, 225
Dentition, 41
Dermal
  condensations, 155
  -epidermal junction, 151

[Dermal]
messages, 157
papillae, 166, 170-71, 178, 179, 229, 230
Dermis, 147, 174, 175, 176, 177, 178, 179, 182, 183
chick, 176, 182, 184
dense, 125
dorsal chick, 147
dorsal mouse, 147
head, 125
mouse, 177, 182
mouse lip, 125
scaleless, 127-28, 134
scutate scale, 136
upper lip, mouse, 147
Development
cranio-facial, 222
limb, 3
cutate scale, stages of, 116
tooth, 31
Dew-claws, 235
Diastema, 221, 224
Differential growth processes, 219
Differentiation, 4, 22, 121, 163, 166, 183
Digits, 16, 17-18, 19
Digitogenesis, amphibian, 218
Distribution pattern, 147
Dolichos biflorus, 138
DNA synthesis, 116

Ectoderm, embryonic, 157
Ectodermal, 39, 115, 127, 163
Ectomesenchyme, 222
Ectopic bones, 189, 202
Ectopic cartilages, 189, 196, 202
Electron microscope, 137
Embryos, 172, 183
mouse, 167
quail, 124
Endodermal, 39, 127

Epidermal
-dermal junction, 127
competence, 156
placodes, 155, 230
feather, 136
morphogenesis of, 136
Epidermis, 163, 174, 175, 176, 178, 179, 180, 182
apteric, 124, 128
backskin, 124, 127, 133
beak, 124
chicken, 163
dissociated, 138
dorsal, 155
lizard, 147
mouse, 176, 182
neutral, 154
reticulate scale, 129
scaleless, 134
Epidermolysis bullosa, 44
Epithelial
cells, 79, 167
collagen, 204
derivatives, 33
glycosaminoglycans, 137, 175, 184, 193
-mesenchymal interactions, 3, 22, 64, 79, 230
morphogenesis, 52, 66
Epithelium, 41, 51, 177, 179
amniotic, 126
bladder, 60
chorionic, 121, 126, 134
corneal, 124
mammary, 34
mandibular, 196, 206
Mullerian, 57
pigmented retinal, 194
salivary, 34
stratified squamous, 94
skin, 28
uterine, 57
vaginal, 57
wing and leg bud, 201

Estrogen receptor, 57
Eudiplopod, 227
Evolution, 21, 221
Evolutionary change, 21, 215
Evolved inductive systems, 224
Expressive events, 174
External ear cartilage, 189
Extracellular, 137, 138
  matrix, 29, 31, 32, 44, 75-
  76, 84, 85, 86, 87, 93-94,
  179
Extraembryonic tissues, 127,
  155, 157

Faculty, 217
Feathers, 115, 136, 147, 220-
  21, 229, 230
  chimaeric, 234
  dermal condensations, 137
  morphogenesis, 138
  placode, 136
  primordia, 138
Features, 217
  fine structural, 129
  neomorphic, 217
  skeletal, 225
Feet, 183
  feathered, 156
Fibrils, 175
Fibrin, 82
Fibroblasts, 86, 174
Fibronectin, 30, 77, 82, 83,
  86, 98, 138, 224, 226, 227
Flightless, 227
Form, 217
  convergent, 218, 225
  of a feature, 217
  fossilizable, 225
  genesis of, 218
  recapitulation of, 215
Fossil record, 216
Functions, 217

Galactosamine, 81
Galliformes, 21

Gaps, 169
Gap junction, 13, 21
Gel electrophoresis, 136
Genes, 131
  expression, 116
  quiescent, 233
Genetic
  drift, 232, 233
  ptilopody, 154
  restriction, 235
Genome
  scaleless, 136
Genotype, 221
Glands, 35, 229
  accessory sexual, 52
  mammary, 33, 34, 35, 41,
  52, 56, 64, 66
  mucous, 171, 172, 177
  prostate, 51, 52, 56, 60-61,
  66
  salivary, 33, 34, 35, 37, 51,
  64
  sebaceous, 221
  sweat, 221
Glandular morphogenesis, 33,
  35, 165, 171, 172, 174, 177,
  178, 183, 184
Glucuronic acid, 81
Glycoproteins, 82, 137, 138,
  184
Glycosaminoglycans, 80, 81,
  86, 137, 175, 184, 193
  epithelial, 204
Gradualistic models, 216
Grafts, 175, 176
Growth, 22, 81
Gruiformes, 21
Gut, 39

Hair, 28, 33, 34, 41, 115, 147,
  165, 176, 178, 229
  buds, 176, 177
  bulb, 170
  follicles, 147, 163, 165, 166,
  174, 176, 182

[Hair]
  matrix, 166, 171, 178, 179
Hamster, cheek pouch, 182,
  183
Hemidesmosome, 106, 167,
  169
Heparin, 82, 86
Heterochronic
  growth, 228
  modulation, 225
Heterodonty, 221, 228
Heterogenetic leg bud recom-
  binants, 156
Heterospecific dermal-
  epidermal recombinations,
  147
Heterotopic recombinations,
  147, 150
  limb bud, 155
Heterotypic, 121, 234
  cell contacts, 171, 183
Hexosamine, 80
Hexuronic acid, 80
Histogenesis, 147
Histological, 118
Homodont teeth, 220, 233
Homospecific recombinations,
  147
Hormone receptors, 52
Hormonal sensitivity, 52
Hyaluronidase, 204
Hydrocortisone, 172
Hyperphalangic, 226

Ichthyosaurs, 226
Indirect immunofluorescent,
  134, 138
Induction
  apical ectodermal ridge, 7,
  9, 10
  lens, 63
  neural, 222
  scale, 63
  stability, 28, 32-33

Inductive
  ability, 136
  capacities, 224
  cues, 134
  signal, 29, 31
Initiation, 147
Inner epidermal surface, 127
Instructive interaction, 10, 166,
  171, 178, 179, 182, 183
Integument, 28, 33, 34, 39, 41,
  42, 44, 157, 220, 221
Integumentary heterogeneity,
  220
Interactions
  epithelial-mesenchymal, 31,
  34, 51, 230
  epithelial-stromal, 66
  instructive, 7, 22, 31, 32-33,
  44, 166, 171, 178, 179, 182,
  183
  permissive, 10, 11, 22, 31,
  171, 190, 200
  second dermal, 170
  tissue, 27, 43
Intercellular matrix, 175, 177

Kangaroo, 220
Keratan sulfate, 3
Keratin, 163, 184
  alpha, 127, 133
  antisera to alpha and beta, 133
  beta, 127, 133, 134, 136
  intermediate, 133
Keratinization, 41, 103, 163
  179, 182
Keratinocyte, 184
Kiwi, 227

Lagomorphs, 229, 235
Lamellar ichthyosis, 41
Lamina densa, 137
Leukoplakia, 68
Lectin, 138
Leg, 11

Limb
  bud, 3, 4, 19, 20, 21, 22,
    192, 233
  cartilages, 189
  chimaeric, 233
  cursorial, 224, 225
  development, 3
  morphogenesis, 222
  recombinant, 11
Lips, glabrous, 221
Lizard epidermis, 147
Lucifer yellow CH, 13

Macroevolutionary phenomena,
    216
Macromutations, 227, 229
Maintenance factor, 7
Mammals, 147, 224
Mandibular
  bones, 199
  epithelium, 196, 205
Mammary carcinogenesis, 68
Matrix
  extracellular, 179
  hair, 166, 171, 178, 179
  intercellular, 175, 177
Mechanistic developmental
    biology, 215-16
Meckel's cartilage, 189, 195,
    196, 205, 220
Membrane bones, 189, 200, 225
Mesenchyme, 115
  capsular, 34
  cells, 167, 175
  condensations, 229-30
  dental, 230
  papillar, 34
  perivibrissal, 231
  specificity, 33
Mesodermal, 127
Mesoglea, 76
Messages, 147
Mice, 175
  embryonic, 175
Microfilaments, 87

Microtubules, 87
Mitotic activity, 202
Moas, 227
Molariform, 233
Molar patterning, 223
Morphogenesis, 35, 52, 57, 68,
    81, 85, 115, 147, 219
  abnormal, 129
  appendage, 147
  epidermal placode, 136
  epithelial, 64
  feather, 138
  glandular, 165, 171, 172, 174,
    177, 178, 183, 184
  limb, 222
  scale placode, 155
  skin, 155
  tooth, 28
Morphogenetic, 223, 235
Mouse, 176
  -chick recombinations, 183
  dermis, 176, 177, 182
  embryos, 167
  epidermis, 176, 182
  limb buds, 192
  lip dermis, 125
  -mouse recombinations, 183
  pelage hair follicles, 147
  -quail recombinations, 234
  vibrissa follicles, 183
  skin, 165
mRNA, 134
Mucopolysaccharide, 166
Mucosubstances, 175, 179
Mucous
  differentiation, 174
  glands, 171, 174, 177
  metaplasia, 163, 165, 174,
    179, 182
  secreting glands, 165, 183
  secretion, 172
Multituberculates, 228
Mutants, 41
  dental, 41
  hairless, 41

[Mutants]
  limbless, 11, 12
  nanomelia, 82
  pupoid fetus, 41
  scaleless, 39, 230, 235

Nails, 229
Nasal capsular cartilage, 189
Natural selection, 217
Neomorphic features, 217
Neoteny, 227
Neural
  crest, 190, 194-95, 205, 222
  induction, 222
Neutral epidermis, 154
Notochord, 193

Odontoblasts, 223, 231
Oligosaccharides, 3
Ontogeny, 215
Optic vessicle, 234
Oral mucosa, 41, 42
Organ culture, 165, 178
Otic
  capsule cartilage, 189, 195,
    206
  epithelium, 195
Outer epidermal surface, 127

Paddle-limbs, 224, 226
Pads, 235
Paleodevelopmental biology,
  233
Papillae
  conjunctival, 118
  dermal, 166, 170, 178, 229,
    230
  scleral, 198-99
Pattern
  distribution, 147
  expression, 223, 225
  formation, 115, 118, 219
Patterning, 44
Patterning properties, 156
Pelycosaurs, 228

Perissodactyls, 225
Perivibrissal mesenchyme, 231
Permissive interactions, 171,
  190, 200
Pharyngeal endoderm, 195, 205
Phenocopies, 227
Phenotype, 134, 221
Phylogeny, 215
Pigmented retinal epithelial, 194
Pinnipedia, 22
Plasma lemma, 106
Plasma membrane, 29
Plasmid (pCSK-12), 134
Pleiotropism, 228
Polar coordinate model, 222
Polydactyly, 226
Polygenic system, 226
Population geneticists, 232
Positional information
  hypothesis, 28, 221, 223, 224,
    230, 233
Postdentary elements, 225
Prepatterning, 222
Procelliformes, 21
Progress zone, 222, 224
Protein synthesis, 62, 184
Proteoglycans, 77, 79, 86, 106,
  137
Psoriasis, 41
Pterygo-quadrate bars, 220
Ptilopody, 150, 156
Punctuated equilibria, 216

Quail embryos, 124
Quiescent genes, 233

Rachis, 234
Reciprocal heterotypic recombinations, 121
Recombinations, 60
  chick-mouse, 178, 183
  heterochronics, 11
  heterogenetic leg bud, 156
  heterospecific dermal-
    epidermal, 147

[Recombinations]
heterotopic, 147, 150
heterotopic, limb bud, 155
heterotypic, 68
homospecific, 147
limb, 11
mouse-mouse, 183
mouse-quail, 234
reciprocal heterotypic, 121
tissue, 52, 64, 175
turtle-chick, 234
xenoplastic, 232
Reductionist models, 221
Regional
origin, 229
specification, 151
specificity, 157, 221
Reptiles, 147
theriodont, 225, 228, 232
Reptilian scale, 118, 147, 229
Responding tissue, 121
Restrictive event, 174
Reticula, 150, 151
Reticulate scale epidermis, 129
Retinoic acid, 151, 172, 175,
176, 177, 178, 182, 183,
230, 231
Retinoids, 172, 174, 177, 178,
183, 184
Retinol, 166, 167, 172, 175,
178
binding protein, 174, 183
Retinyl acetate, 172, 174, 175,
184
Ridge (see apical ectodermal
ridge)
Rodents, 229

Saber-tooth tigers, 225
Sauropterygians, 226
Scaled integument, 220
Scale-inducing properties, 154
Scaleless, 230, 235
anterior metatarsal dermis,
127-28

[Scaleless]
cell surfaces, 138
claws, 136
dermis, 134
epidermis, 134
gene, 133
genome, 136
high line, 136
reticulate scales, 133
Scales, 154, 155, 183, 221
avian
scutate, 116, 118, 134, 136
scutellate, 118
spur, 118
reticulate, 118, 129, 133
feathered, 150
reptilian, 118, 147, 229
S-carboxymethylated polypeptides,
131
Scleral
bones, 198, 199
cartilage, 189, 194, 205
ossicles, 118
papilla, 198
Scuta, 149, 151
Scutate scale dermis, 136
Scutella, 149, 151
SDS-PAGE, 131
Sea-turtles, 226
Sebaceous glands, 221
Second dermal interaction, 170
Second specific dermal message,
176
Selection, 226, 232, 235
pressure, 217
Seminal vesicle, 51
Sex steroids, 52
Silkies, 150
Skeletal features, 225
Skin, 27, 41, 106, 115, 147,
163, 170, 172, 174, 178,
179, 183
mouse, 165
thigh, 136
Skull bones, 189, 197

Somatopleura, 124, 125
Source-sink diffusion models, 222
Species, 121
Spurs, 130, 133
Stork, 220
Stratum
  basal, 134
  granulosum, 179
  intermedium, 134
  corneum, 134, 179
Structural organization, 147
Stylopod, 220
Subperidermal layer, 136
Subridge mesoderm, 6-7
Sulphated
  acidic glycosaminoglycans, 179
  proteoglycans, 137
Sweat glands, 221
Syndactylous, 226
Synthetic evolutionary theory, 216

Talpid, 226
Tarsometatarsal, 148
  dermis, 150, 155
Teeth
  chicken, 231
  homodont, 220, 232
  mammalian, 33, 34, 41, 220
Temporal and spatial, 137
Teratocarcinomas, 197
Testicular feminization, 52
Theriodont reptiles, 225, 228, 232
Thigh skin, 136
Tissue interactions, 33, 44
  instructive, 7, 22, 31, 32, 166, 171, 178, 179, 182, 183
  permissive, 10, 11, 22, 32, 171, 190, 200
Tissue recombination, 175
Tissue-specific keratin polypeptides, 131

Tongue, grooved, 42
Tooth, 31-32
Tooth and limb evolution, 219
Tooth germ, 35, 41
Transformation series, 226
Transglutaminase, 84
Transient condensations, 231
Tricondodonts, 228
Trigeminal
  ganglion, 223
  innervation, 223
Triturus, 234
Trypsin, 137, 175, 176
Tumor cartilages, 189, 196
Turtle-chick recombinations, 234

Ubiquitous extracellular protein, 224
Ulex europeus, 138
Ultrastructural, 121
Ungulate, 228
Urogenital
  epithelia, 52
  glands, 52
  mesenchyme, 55, 64, 66
  morphogenesis, 51
  sinus mesenchyme, 51, 60, 63
  tract, 33, 68
Uterus, 57, 58

Vagina, 57, 58, 66
Vertebral cartilage, 33, 190
Vertebrates, 115
Vestigial dental laminae, 232
Vestigiality, 225
Vibrissae, 33, 34, 41, 42, 147, 221
  follicles, 165, 167, 171, 172, 174, 175, 179, 182, 183, 184
  papillar cells, 231
Vitamin A, 34, 163, 165, 166, 167, 171, 172, 174, 175, 177, 178, 179, 180, 182, 183, 184
  fluorescence, 174

Webbing, 226
Wing and leg bud epithelia, 201
Wingless, 228
Wound healing, 7, 8, 83

Xenoplastic recombinations, 232

X-ray
  diffraction, 121, 136
  irradiation, 156

Zeuogopod, 220